STD Bus Interfacing

CHRISTOPHER A. TITUS,

JONATHAN A. TITUS,

and

DAVID G. LARSEN

Order From
GROUP TECHNOLOGY, LTD.
P.O. Box 87
Check, VA 24072
(703) 651-3153

Copyright © 1982 by Christopher A. Titus, Jonathan A. Titus, and David G. Larsen

FIRST EDITION
FIRST PRINTING—1982

All rights reserved. No part of this book shall be reproduced, stored in a retrieval system, or transmitted by any means, electronic, mechanical, photocopying, recording, or otherwise, without written permission from the publisher. No patent liability is assumed with respect to the use of the information contained herein. While every precaution has been taken in the preparation of this book, the publisher assumes no responsibility for errors or omissions. Neither is any liability assumed for damages resulting from the use of the information contained herein.

International Standard Book Number: 0-672-21888-7
Library of Congress Catalog Card Number: 82-50652

Edited by: *Frank Speights*
Illustrated by: *David K. Cripe*

Printed in the United States of America.

Preface

Ever since the first microcomputers became available in 1974, users have been interested in the microcomputer's bus. The *bus* is the collection of signals that the microprocessor uses to communicate with and control the memory and the various peripherals. In general, as new microprocessors of central-processing units (CPUs) became available from the semiconductor manufacturer, the microcomputer manufacturers would design and sell a microcomputer based on this new chip. The problem was (and still is) that each manufacturer thought they knew best what the user wanted and needed, so each microcomputer was designed with different sized cards, different power-supply voltages, different signals, and different arrangements of signals on the bus. Thus, at this time, we have the KIM, TRS-80, Atari, Apple, Heath, and OSI buses plus the S-100 bus, the IEEE 488 bus, the Multibus, and the Versabus, to name just a few. In addition, there is also the *STD bus,* which is the basis for all of the discussions presented in this book.

What's different about the STD bus? Well, when originally conceived, the STD bus was supposed to be a universal arrangement of signals that any, or just about any, microprocessor could use. Thus, the STD bus started out being *processor independent,* which most of the other bus systems were not. This means that it should be possible to design new STD bus CPU cards as new CPU chips become available, and these new cards should work in old systems, simply because all of the signals and pin assignments on the edge connector were previously defined for a general-purpose CPU chip.

Unfortunately, this is the ideal case. It soon became apparent that as new CPU chips were designed, they no longer could generate some of the bus signals. On the other hand, some of the CPU chips didn't need to generate some of the bus signals, but did need to put other signals on the bus. Thus, at this time, most manufacturers make some cards that can be used in *any* STD bus system (regardless of the CPU) along with cards that can only be used with a specific CPU chip. Therefore, some cards are specified to be STD bus compatible, while some cards are only STD Z-80, STD 6800, STD 8085, or STD 6502 compatible. Thus, what was originally conceived as a standard or universal bus has now become, in some cases, very processor specific.

There is also a second "bus" at the other end of the STD bus cards, which is used with interrupt and direct-memory access (DMA) devices. However, because of the nature of these signals, there shouldn't be any incompatibility problems.

Even though some cards are processor specific, it is still relatively easy to put together a system that works. According to one manufacturer, customers can purchase between 80-90 percent of their cards from the manufacturers of STD bus cards. The remaining 10-20 percent of the cards have to be designed and built by the end user. In general, those cards requiring user action are used to interface some type of nonstandard peripheral device to the STD bus computer. Thus, in this book, interfacing will be described, along with the chips that are used to transfer information between a peripheral and the CPU. Also, the techniques used to decode addresses or assign addresses to peripherals, data transfer and control-signal timing, and input/output (I/O) software will be discussed. We will not describe how memory or CPU cards are designed because you can probably buy just what you need.

Unfortunately, every topic of interest cannot be covered in depth in this book. Therefore, you may want to refer to some of the other Blacksburg series of books:

1. 8080/8085 Software Design, Books 1 & 2.
2. Microcomputer—Analog Converter Software and Hardware Interfacing.
3. 6502 Software Design.
4. 6809 Microcomputer Programming and Interfacing, With Experiments.
5. 8085A Cookbook.
6. How to Program and Interface the 6800.

7. Advanced 6502 Interfacing.
8. Microcomputer Design and Troubleshooting.
9. Programming and Interfacing the 6502, With Experiments.

One final note. In many of the chapters, we have included assembly language software listings. The 8085 and Z-80 programs were edited and assembled on a microcomputer, but the 6800 and 6502 programs were assembled by hand. Thus, the instructions in these programs may not be in exactly the right format for your assembler. Since the listings were typed by hand, we may have inadvertently added some bugs to the programs. In Chapter 6, we had to simulate the I/O cards with discrete input and output ports, so the software may not be exactly what is needed to drive the I/O card.

During the course of writing this book, we received a lot of help from a number of manufacturers and their representatives. They supplied us with microcomputers, technical manuals, and answered a lot of questions over the phone. Special thanks to:

Bob Burckle, Cheryl Gartenberg, Bill Bradley, and John Nichols—Mostek Corporation
Dick Thomas—Pro-Log Corporation
Rick Miller—Rep-Tron
Claudia Krueck, Ken Boyette, and Randy Wrenn—Matrix Corp.
Al Haun—Analog Devices, Inc.
Sandra Lawson—Enlode, Inc.
Bob Johnson—Atec, Inc.
George Perrine—Enterprise Systems Corporation
Eric Miller—Miller Technology

We hope that you enjoy this book, and are able to put some of the material to good use. By the way, don't forget to fill out and send in the Reader Service Card located in the back of the book.

"The Blacksburg Group"
DAVID G. LARSEN, JONATHAN A. TITUS,
AND CHRISTOPHER A. TITUS

Contents

CHAPTER 1 — 11

WHAT IS THE STD BUS?

Other Control Signals—Physical Standards—Why Use the STD Bus?—STD Bus Processors—Memory—Read-Only Memories—Read/Write Memory—Control Signal Generation—Memory Maps—I/O Devices—Memory-Mapped I/O—Accumulator I/O—CPU Compatibility—I/O Interface Compatibility—Data Transfer Timing—Chip Incompatibility—Nonstandard Signals—Conclusion

CHAPTER 2 — 41

I/O DEVICE ADDRESSING

Address and Control Signals—Device Addressing—Using Gates for Address Decoding—Using Decoders—Larger Decoders—Using Comparators—Memory-Mapped I/O—Using PROMs—Conclusion—References

CHAPTER 3 — 81

OUTPUT PORT INTERFACING

Output Timing—Latches in Output Ports—A Traffic Light Controller—LED Displays—Digital-to-Analog Converters—Data

Displays—Other DAC Considerations—I/O Chips—Memory-Mapped Output Ports—Conclusion—References

CHAPTER 4 — 123

INPUT PORTS

Designing Input Ports—An ASCII Keyboard Interface—Flags—Another Keyboard Interface—An Analog-to-Digital Converter Interface—A Simple Logic Tester—Memory-Mapped Input Ports—Conclusion

CHAPTER 5 — 161

INTERRUPTS AND DIRECT-MEMORY ACCESS

Basic Interrupt Operation—STD Bus Interrupts—The 8085—The 8088—The Z-80 and NSC800—Serial Priority—The 6800, 6809, and 6502—6809 Improvements—An Interrupt Review—Interrupt Software—Interrupts and the Stack—Interrupt Timing—Direct-Memory Access—Requesting the Bus—Direct-Memory Access Controllers—DMA Software—Conclusion

CHAPTER 6 — 195

GENERAL-INTEREST INTERFACE CARDS

The Mostek DIOB/DIOP—The Enlode 214 Display System—The Analog Devices RTI-1260—The Atec 710 Thumbwheel Switch Interface—The Matrix 7911 Stepper-Motor Controller—The Pro-Log 7304 Dual UART Card

APPENDIX A — 259

THE STD BUS STANDARD

Introduction—Organization and Functional Specifications (With Pin Definitions)—Electrical Specifications—Mechanical Specifications

APPENDIX B ——————————————— **273**

VOLTAGE INPUT CONFIGURATIONS

Input Multiplexer Guidelines—Analog Input Multiplexer

APPENDIX C ——————————————— **277**

INDEX OF STD BUS MANUFACTURERS

INDEX ——————————————————— **282**

What Is the STD Bus?
Chapter 1

What is the STD bus? Very simply, the STD bus is a collection of 56 signals that are used by a microprocessor to communicate with memory and peripheral devices. The signals on the bus are also used by some direct-memory access (DMA) devices, such as a high-speed "Winchester" disk drive, to access memory directly without going through the microprocessor chip. The signals in the STD bus consist of address, data, and control signals, along with 10 pins or conductors that are dedicated to power. The physical arrangement of these signals is shown in Table 1-1.

From Table 1-1, you can see that there are 16 signals labeled A0–A15 and there are also signals labeled D0–D7. The address signals (A0–A15) are used by the microprocessor chip on the central-processor unit (CPU) card to address memory and peripherals. Since these signals can assume one of two possible states, there are 2^{16} or 65,536 unique combinations of 1s and 0s (or 65,536 different addresses) that can be present on these 16 address lines. In fact, most of the processors that are available for use on an STD bus can address all 65,536 memory locations, so they can generate every one of these 65,536 unique memory addresses. Of course, many systems won't need this much memory, which is quite acceptable. Processors that are available on STD bus CPU cards, and which can address this much memory, include the 8085, Z-80, 6502, 6800, NSC800, and the 6809.

Of course, not only must the CPU be capable of addressing memory, it must also have the ability to transfer information between itself and memory. This information is transferred on the data bus lines

Table 1-1. STD Bus Signals and Their Location

Signal	Component Side				Circuit Side			
	Pin	Mnemonic	Signal Flow	Description	Pin	Mnemonic	Signal Flow	Description
Logic Power Bus	1	+5V DC	In	Logic Power (bused)	2	+5V DC	In	Logic Power (bused)
	3	GND	In	Logic Ground (bused)	4	GND	In	Logic Ground (bused)
	5	VBB #1	In	Logic Bias #1 (−5V)	6	VBB #2	In	Logic Bias #2 (−5V)
Data Bus	7	D3	In/Out	Low-Order Data Bus	8	D7	In/Out	High-Order Data Bus
	9	D2	In/Out	Low-Order Data Bus	10	D6	In/Out	High-Order Data Bus
	11	D1	In/Out	Low-Order Data Bus	12	D5	In/Out	High-Order Data Bus
	13	D0	In/Out	Low-Order Data Bus	14	D4	In/Out	High-Order Data Bus
Address Bus	15	A7	Out	Low-Order Address Bus	16	A15	Out	High-Order Address Bus
	17	A6	Out	Low-Order Address Bus	18	A14	Out	High-Order Address Bus
	19	A5	Out	Low-Order Address Bus	20	A13	Out	High-Order Address Bus
	21	A4	Out	Low-Order Address Bus	22	A12	Out	High-Order Address Bus
	23	A3	Out	Low-Order Address Bus	24	A11	Out	High-Order Address Bus
	25	A2	Out	Low-Order Address Bus	26	A10	Out	High-Order Address Bus
	27	A1	Out	Low-Order Address Bus	28	A9	Out	High-Order Address Bus
	29	A0	Out	Low-Order Address Bus	30	A8	Out	High-Order Address Bus

Table 1-1. Cont.

Signal	Component Side				Circuit Side		
	Pin	Mnemonic	Signal Flow	Description	Pin	Mnemonic	Signal Flow	Description
Control Bus	31	WR*	Out	Write to Memory or I/O	32	RD*	Out	Read Memory or I/O
	33	IORQ*	Out	I/O Address Select	34	MEMRQ*	Out	Memory Address Select
	35	IOEXP	In/Out	I/O Expansion	36	MEMEX	In/Out	Memory Expansion
	37	REFRESH*	Out	Refresh Timing	38	MCSYNC*	Out	CPU Machine Cycle Sync
	39	STATUS 1*	Out	CPU Status	40	STATUS 0*	Out	CPU Status
	41	BUSAK*	Out	Bus Acknowledge	42	BUSRQ*	In	Bus Request
	43	INTAK*	Out	Interrupt Acknowledge	44	INTRQ*	In	Interrupt Request
	45	WAITRQ*	In	Wait Request	46	NMIRQ*	In	Nonmaskable Interrupt
	47	SYSRESET*	Out	System Reset	48	PBRESET*	In	Push-Button Reset
	49	CLOCK*	Out	Clock from Processor	50	CNTRL*	In	AUX Timing
	51	PCO	Out	Priority Chain Out	52	PCI	In	Priority Chain In
Auxiliary Power Bus	53	AUX GND	In	AUX Ground (bused)	54	AUX GND	In	AUX Ground (bused)
	55	AUX +V	In	AUX Positive (+12V DC)	56	AUX −V	In	AUX Negative (−12V DC)

*Low-level active indicator

Courtesy Pro-Log Corp.

D0–D7. Since these signals can also assume one of two possible states, there are 2^8 or 256 different combinations of 1s and 0s that can be present on these lines. Of course, this does not mean that STD bus computers are limited to operating on data that have values between 0 and 255. In fact, data values between 0 and 65,535, and even larger numbers, can be processed quickly and efficiently on one of these microcomputers.

At this point, the processor can address memory and use the data bus to transfer information. However, there has to be at least one additional signal. This signal indicates the direction of data flow on the data bus. For instance, if the microprocessor needs to read information from memory, the microprocessor has to generate a signal which indicates that this type of operation is taking place. Also, if the processor needs to write information into memory, this has to be indicated to the memory by some signal. This signal is required since the data bus can only be used for one operation at a time. It will allow either Read, Write, or *neither* Read or Write.

With some microprocessors, this process requires just one signal. If the signal is a logic 1, then a read operation is taking place, and if the signal is a logic 0, a write operation is taking place. Unfortunately, there are also some microprocessors that use more than one signal to indicate that either a read or write operation is taking place. Thus, when the STD bus was defined by Mostek and Pro-Log Corporation, four signals were assigned to indicate the read or write status of the microprocessor: MEMRQ*, IORQ*, RD*, and WR*. As you may already know, when the name of a signal is followed by an asterisk, the signal is active in the logic 0 state (a voltage near ground). Since address and data signals are active in both states, no asterisks are used after these signal names.

In general, the four control signals (MEMRQ*, IORQ*, RD*, and WR*) are all that are needed to indicate to both peripherals and memory whether a read or write operation is taking place. Of course, there is also another possibility. That is when *neither* a read or a write operation is taking place. This would occur if the processor is moving data around inside the CPU chip or is calculating the result of some mathematical operation.

Even though there are 16 possible combinations of 1s and 0s for these four control signals, there are really four combinations that we are interested in and which normally occur in STD bus systems. These combinations are summarized in Table 1-2.

As you can see, we are only interested in the combinations where

Table 1-2. Useful Combinations of RD*, WR*, IORQ*, and MEMRQ*

RD*	WR*	IORQ*	MEMRQ*	Operation
0	1	0	1	Input/output (I/O) read
1	0	0	1	Input/output (I/O) write
0	1	1	0	Memory read
1	0	1	0	Memory write
1	1	1	1	Internal operation

either RD* or WR*, along with either IORQ* or MEMRQ*, are generated. With RD* a logic 0, a read operation would be taking place, and with WR* a logic 0, a write operation would be taking place. If MEMRQ* is a logic 0, a memory operation would be taking place and if IORQ* is a logic 0, an input/output (I/O) operation would be taking place. Thus, the computer may be performing a memory read, a memory write, an I/O read, or an I/O write. As was mentioned previously, it is also possible for the microprocessor to be doing none of these operations, particularly when it is performing some internal operation.

OTHER CONTROL SIGNALS

By referring back to Table 1-1, you can see that there are 18 other control signals that have not been discussed. These signals are either used to give us more detailed information about what the processor is doing or are used to control the processor.

For instance, the MCSYNC*, STATUS 1*, and STATUS 0* signals are generated by the CPU card to synchronize the transfer of data between itself and some peripheral device. Ordinarily, these signals are not used.

The REFRESH* signal is usually only generated by Z-80- or NSC800-based CPU cards. It is used, along with an address on the address bus, to refresh dynamic read/write memory (RAM). In general, because STD bus cards are fairly small (4.5 inches × 6.5 inches), dynamic read/write memories are only used in Z-80- or NSC800-based systems. This is because the refresh functions that are performed by these CPU chips would have to be performed by discrete logic in other CPU systems.

Unfortunately, as you will see a little later in this chapter, these signals are really not standard at all. Depending on the CPU card, these signals may be assigned a totally different function. Thus, in a 6502-based system, no signal is present on REFRESH*, Ø2* is present on the MCSYNC* line, SYNC* is present on the STATUS 1* line, and R/W* is present on the STATUS 0* line.

The MEMEX* and IOEXP* signals are used for memory and I/O expansion, respectively, and are not used in small systems and are not used very frequently. One of the few times that MEMEX* is used is when there are two sections of memory that have the *same address* and, thus, only one section of memory may be enabled at a time.

As we mentioned previously, it is possible to use direct-memory access (DMA) devices with many STD bus microcomputers. A DMA device can access memory directly without having to transfer information to memory through the CPU chip. Thus, a high-speed disk might write information directly into memory or read information directly from memory. While this type of transfer is going on, the DMA device uses the address, data, and control buses. *Thus, the microprocessor on the CPU card cannot read instructions from memory or execute them while a DMA transfer is taking place.* The BUSRQ* and BUSAK* signals are used by DMA devices to get control of the address, data, and control buses from the microprocessor. The BUSRQ* signal is generated by the DMA device to request use of the buses. Then, when the microprocessor is ready to give up control of the buses to the DMA device, it generates the BUSAK* signal. In simple systems, these signals are not used. One example of where these signals are used is on the Mostek MDX-FLP floppy disk drive interface card.

There are also five interrupt signals that peripherals can use to "get the attention of" the CPU card. These signals are INTRQ*, INTAK*, NMIRQ*, PCI, and PCO. The PCI and PCO signals are fairly specific to the Z-80 microprocessor, so other types of processors either won't use them or will use another interrupt technique. (This technique will be discussed in Chapter 5.)

The WAITRQ* signal Is generated by memory and by peripherals when they need a short amount of additional time (typically a few μsec) to respond to the signals generated by the CPU. Thus, this signal has the effect of telling the CPU to "wait." This signal is rarely used today because the memory and peripheral chips currently available are very fast. They can respond to the microprocessor's requests in from 10 to 450 ns.

Finally, there are two reset signals (SYSRESET* and PBRESET*) and two clock signals (CLOCK* and CTRL*). The SYSRESET* signal is generated by the CPU card whenever the CPU chip is reset. Usually, there are two ways that this can be accomplished—either by using an external push button or when power is first applied. The PBRESET* signal is actually generated by the external push button.

The clock signal of greatest interest to most users is CLOCK*, because this is the clock that is used to clock the CPU chip. Thus, the higher the frequency of this signal, the faster that the microprocessor can perform a task. This signal is usually generated by logic on the CPU card and is then buffered before it is brought out to the bus. The CNTRL* signal is another clock signal generated by the CPU card, but it may be a multiple of the basic processor's clock (CLOCK*). Thus, there is no real standard for this signal.

Finally, there are two conductors in the bus that are dedicated to +5 volts and two that are dedicated to ground. Of course, this voltage is used to power most of the chips in the STD bus system. There are also two pins for a −5 V supply, if you need one, along with four pins for an auxiliary ± supply. Currently, there is no standard concerning what these voltages should be, although, in most systems, they will be either ±12 V or ±15 V.

PHYSICAL STANDARDS

Not only were the pins for the individual signals defined but the size of the cards was also specified. Thus, most STD bus cards are 4.5 inches wide and 6.5 inches long (11.43 cm × 16.51 cm), with a 56-conductor edge connector located along one of the 4.5-inch sides. In general, if a card doesn't fit these dimensions, it is because the card is longer than the 6.5 inches specified. This often happens when the manufacturer needs to provide the user access to something on the card. A typical example would be an EPROM (Erasable Programmable Read-Only Memory) programmer card, where the EPROM that is to be programmed is inserted into a socket that sticks out above the other cards in the system. If this card were only 6.5 inches long, the user would have to remove the card from the system each time an EPROM had to be programmed. The more that this was done, the greater would be the chance of having a problem with the system.

The physical dimensions of the card also specify that a notch will be present in all cards between pins 25 and 27 (26 and 28 on the opposite side of the card). This prevents cards from being inserted into a system backwards (since the edge connector is right in the center of the card). Older cards may not have this notch in them, so be careful when taking cards out and when putting them back into a system. Rather than discuss any more of the physical and electrical standards, we want you to refer to the complete STD bus standard located in Appendix A.

WHY USE THE STD BUS?

At this time, there are over 20 bus "standards." Thus, the user could spend many months evaluating the different buses and the CPU cards that are available to them. However, there are really only three general-purpose buses available which can be used with many different CPUs: the S-100 bus, the Multibus, and the STD bus.

One basic problem of the S-100 bus is that it was not well thought out. Many of the signals that were originally placed on the bus were specific to the 8080, which made interfaces and memory boards more complex than they had to be. This has been improved and there are a number of different CPUs available for the S-100 bus (8080, 8085, Z-80, 8088, 6502, and 9900). The S-100 is also being considered as a standard by the Institute for Electrical and Electronics Engineers (IEEE). Another problem with this bus is that, originally, a number of pins are left undefined, so many manufacturers used them for special functions when they designed boards. Thus, some memories won't work with some processors, some peripherals won't work with some memories, etc.

On the other hand, the Multibus is rigidly defined, and CPU cards for the 8080, 8085, 8088, 8086, 68000, and Z8001/2 are available. However, the cards are large and complex, they have two edge connectors at one end, and they are expensive. For many applications, they have too many features.

Finally, the STD bus is fairly well defined and the cards are relatively small and simple. There are some incompatibilities between cards with respect to the MCSYNC*, STATUS 1*, STATUS 0*, and REFRESH* signals. In general, however, the manufacturers have specified which CPU cards can be used with their peripheral cards, so the number of incompatibility problems should be small. The cards are also relatively low cost, which means that you can get started fairly easily.

STD BUS PROCESSORS

Currently, there are eight different CPU chips that are available on STD bus cards: the 8085, Z-80, NSC800, 6800, 6809, 6809E, 6502, and the 8088. There is really only one requirement that a CPU chip must meet in order to be used with the STD bus, and that is that it use an 8-bit data bus. Some users might expect that another requirement would be that the CPU can address a maximum of 65,536 (64K) memory locations, but in the case of the 8088, it can address 1,048,576 memory locations! However, in order to have this much memory in an STD bus system, the user would need some specially designed memory cards. At this time, there are a number of other processors that meet this single requirement, including the 9980 and the 9995 (TI), the Z800 (Zilog), the 1802 (RCA), and the 6801, 6802, and 6803 (Motorola).

It is hard to imagine that CPU cards using the 8086 (Intel), 68000 (Motorola), Z8001/2 (Zilog), or 16032 (National) processors will be available in the near future, if ever. The reason that these chips will probably never be available on STD bus cards is that they all use a *16-bit data bus to communicate with memory and with peripheral devices*. Since the STD bus only contains eight lines in the data bus, these processors would not be able to fetch instructions from memory or write information out to the peripherals. Because of the cost and complexity involved, it is not very likely that the 16-bit data bus will be multiplexed, 8 bits at a time, onto the 8-bit data bus.

As we mentioned before, there are basic differences between the CPU cards designed around these eight different CPU chips. In order to reduce the number of incompatibility problems, four signals in the STD bus have been redefined, depending on the CPU card being used in the system. These signals, and their functions, are summarized in Table 1-3.

From this table, you can see that only Z-80- and NSC800-based CPU cards generate the REFRESH* signal (pin 37). Thus, dynamic read/write (R/W) memory is usually used just with these two processors. If you have a peripheral card that was designed to use the SO* signal generated by the 8085 (pin 40), it probably won't work in a 6502-based system, because the 6502 would be placing its R/W* signal on this line. Since these signals have different timing and functions, the peripheral card will not work properly.

The manufacturers realized that this could cause a lot of headaches if they weren't careful (which they were), so subsets of the STD bus

Table 1-3. Processor-Specific Signals on the STD Bus

Chip Type	REFRESH* Pin 37	MCSYNC* Pin 38	STATUS 1* Pin 39	STATUS 0* Pin 40
8080	—	SYNC*	M1*	—
8085	—	ALE*	S1*	S0*
NSC800	REFRESH*	ALE*	S1*	S0*
8088	—	ALE*	DT/R*	SSO*
Z-80	REFRESH*	(RD*+WR*+INTAK*)	M1*	—
6800	—	Ø2*	VMA*	R/W*
6809	—	EOUT* (Ø2*)	—	R/W*
6089E	—	EOUT* (Ø2*)	LIC*	R/W*
6502	—	Ø2*	SYNC*	R/W*

*   Low-level active *Courtesy Pro-Log Corporation*
—   Not used
R/W*   Read high, write low
DT/R*   Data transmit high, receive low

were defined. Therefore, if a card is said to be *STD bus compatible,* the manufacturer is saying that the card can be used in *any* STD bus system. If a card can only be used with a specific CPU chip, then the card will be called "STD Z-80" or "STD 6800" compatible. Thus, make sure that you have read all of the specifications before purchasing a card!

The main reason for the incompatibility between peripheral and CPU cards is due to the MOS/LSI peripheral chips that have been used in the design of the peripheral card. A peripheral chip designed to be used with the Z-80 microprocessor probably won't work with the 6502 or the 6800 processor and a chip designed to work with the 6800 microprocessor probably won't work with the 8085. Of course, the chips *can be made* to work with different CPU chips, but this will increase the cost and complexity of the card, so it really isn't worth the effort.

MEMORY

As you probably already know, there must be both memory *and* peripherals in a computer system in order for it to perform a useful task. Memory is used to store instructions, data, temporary results, etc., and peripherals are used by the microcomputer to communicate with the "outside" world. Typically, this combination of memory and peripherals might be used to get data from a keyboard or from an analog-to-digital converter, cause the processing of this information, and, then, output the result to LED displays, a CRT, or even another computer!

Based upon the STD bus standard, there are 16 address lines (A0–A15) which the microcomputer can use to address one memory location out of a possible 65,536 memory locations. Using the notation A0–A15, A0 is the least-significant address bit and A15 is the most-significant address bit. Thus, A0 has a significance of 0 or 1, and A15 has a significance of 0 or 32,768. If all of the address signals or bits are 0, then memory location 0 is being addressed. If all of the address bits are 1s, then memory location 32,768 + 16,384 + 8,192 + . . . + 4 + 2 + 1, or 65,535, is being addressed. Of course, to actually transfer information between the microprocessor and memory, not only is an address required, but also the proper combination of RD*, WR*, and MEMRQ*.

What types of memory are actually used in a microcomputer system? There are actually two basic types: read-only memory (ROM) and read/write (R/W) memory. Typically, the memory chips used in STD bus systems contain 1024 4-bit words, 4096 1-bit words, 16,384 1-bit words, or 2048 8-bit words. If the microprocessor uses an 8-bit data bus to transfer 8-bit values between itself and the memory and peripherals, how can 1-bit "wide" memories be used? In order to use this type of memory, eight chips would be required, with each chip contributing 1 bit to the 8-bit word. In another system, two 1024 4-bit memories might be used to form 1024 8-bit words.

Once some combination of memory chips has been put together to form 8-bit words, how does the microcomputer select one set of memory chips from among all of the others? If a system contains 65,536 memory locations, based on 1024 × 4 chips, there would be 64 2-chip "sets." The only way that the microprocessor can *select* one set out of 64 is by generating *unique* memory addresses for each memory location. Additional hardware also has to be added to the memory cards to use this unique address to select one set of chips. Typically, devices such as decoders and comparators are used to perform this electrical selection. In Chapter 2, some actual circuits that perform this function will be discussed. The pinouts for two typical memory chips are shown in Fig. 1-1.

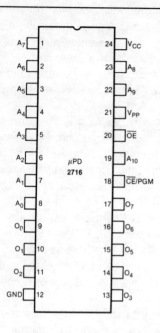

(A) µPD2716

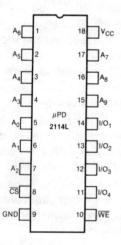

(B) µPD2114L

FIGURE 1-1

Pin assignments for two typical memory chips. *(Courtesy NEC Microcomputers, Inc.)*

READ-ONLY MEMORIES

Read-only memories are used to store programs and data. The important distinction between ROM and R/W memory is the fact that the information in the ROM will stay there, even when the power is removed from the chip. The information stored in R/W memory is *volatile,* because this information will be lost once power is removed.

Read-only memories are available in sizes from 32×4 up to $16,384 \times 8$. Of course, the size, type, and speed of the part (how fast you can get information out of the chip) will determine its cost. There are three basic types of ROMs that will be discussed—mask ROM, programmable read-only memory (PROM), and erasable programmable read-only memory (EPROM).

Mask ROM

The mask ROM can only be programmed by the semiconductor manufacturer while the integrated circuit is being made. This involves a considerable amount of effort, so the manufacturer usually charges the customer a $1000–$3000 "mask charge." In addition, an order for at least 250 of the mask ROMs must usually be placed. The most popular types of mask ROMs have configurations of 2048×8, 4096×8 and 8192×8. These devices may cost between $5.00 and $25.00, depending on the quantity and number of words in the chip. The important points to remember about a mask ROM is that you must supply to the chip manufacturer the information that is to be placed in the ROM, and if there is a mistake in this information, there will be no way that you can correct it once the chips have been made.

Programmable ROM

The programmable ROM, or PROM, is a semiconductor device that contains a number of small electrical fuses. To program the device, *the user selectively blows out the fuses.* An illustration of both a blown fuse and a fuse that is intact can be seen in Fig. 1-2. In many devices, if the fuse is intact, it represents a logic 1. If it is blown, it represents a logic 0. To blow out a fuse, the user has to supply a high-voltage high-current pulse (20–30 V at 20–50 mA) to one of the PROM outputs. Unfortunately, a bit within the PROM can only be programmed (blown) once, so they are not as popular as EPROMs or mask ROMs. For instance, to program 10011101 into one of the PROM's memory locations, you would apply a programming pulse to the D1, D5, and D6 outputs of the PROM. Suppose, however, that after doing this, you realize that a 10111101 should have been programmed into the PROM. This means that D5 should have been a logic 1, rather than a logic 0. Since there is no way for you to replace a blown fuse in the chip, it would have to be thrown away, and a new PROM would have to be programmed with the correct information.

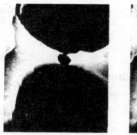

FIGURE 1-2

A fuse that has been blown and a fuse that is intact in a PROM. *(Courtesy Intel Corp.)*

PROMs are available in configurations from 32 × 4 up to 2048 × 8. However, the cost of these devices is in the range of $3.00 to $75.00 per device. Thus, they are not used in microcomputer systems to store long programs; however, they might be used to store a very simple "bootstrap loader." In fact, in many applications, they are not used as a read-only memory but as a programmable gate, as you will see in Chapter 2.

Erasable Programmable ROM

The most popular ROM is the erasable programmable read-only memory (EPROM) invented by the Intel Corporation. This type of ROM can be programmed by the user and then, if the content of the EPROM (data, instructions, etc.) is found to be incorrect, *the content of the EPROM can be erased by exposing it to an ultraviolet (uv) light.* To have this erase feature, EPROMs have a transparent quartz window installed in them so that the ultraviolet light can actually illuminate the silicon IC chip inside the EPROM. Different EPROMs require different uv light intensities and different exposure times in order to be erased. You should refer to the manufacturer's literature for programming, erasure, and electrical characteristics. It is recommended that small opaque labels be placed over the quartz windows of all EPROMs to prevent an accidental erasure due to sunlight or fluorescent lights. The labels also give you the capability of labeling the EPROMs so that you know what program or version of a program is contained in the EPROM.

Surprisingly enough, there are only four different EPROMs that are popular, so this makes your choice of device that much easier to make. Some STD bus memory cards are designed to use any of these devices, while other boards can only use the one type that the manufacturer specifies. These EPROMs are listed in Table 1-4, along with the number of words that they contain and their power-supply requirements.

Table 1-4. Size and Power Requirements of Typical EPROMs

Name	Configuration	Power Supplies
2708	1K × 8	+5, −5, +12
2716	2K × 8	+5
2732	4K × 8	+5
2764	8K × 8	+5

READ/WRITE MEMORY

There are just two different types of read/write memory that are available—static and dynamic. Static R/W memory will retain any information that is in it, as long as power is supplied to the device. However, in order to retain information in dynamic R/W memories, power must be applied *and the memories must be refreshed every 2 to 4 msec.* This refresh circuitry has been built into the Z-80 and the NSC800 CPU chips, so very little additional circuitry is required to use dynamic R/W memories with these processors. On the other hand, a considerable amount of additional circuitry is required to use dynamic R/W memories with the 8085, 8088, 6800, 6809, and 6502 chips, so users of these processors generally use just static R/W memory chips.

Although there is a large number of R/W memory cards to choose from, the STD bus user usually sticks to cards that are commercially available rather than designing his own. It really does not matter which chips are used on a board as long as the distinction between static and dynamic is made.

CONTROL SIGNAL GENERATION

Regardless of which board or boards contain memory ICs, the proper address, data, and control signals have to be present at each chip. Thus, at least part of the address bus must be wired to the chips, probably A0 through A9, or A10 or A11. The data bus must be wired to the chip or chips (D0–D7), along with some combination of RD*, WR*, and MEMRQ*. From Fig. 1-1, you can see the pins that some of these signals are wired to. As you might guess, the address signals are wired to the address pins and the data signals are wired to the data pins, but what do we do with the three control signals? *Some combination of the control signals is gated together with the remaining address signals, and the resulting logic Zero signal is used to enable or select a chip or set of chips.* Thus, not only must RD* and MEMRQ* both be logic 0s when reading information from an EPROM, but the other address bits must be in some user- or manufacturer-defined state, in order to select or enable the chip. Only at this time will the chip place information on the data bus.

Assuming that a 2716 EPROM is present in the system, the EPROM might only be enabled when RD* and MEMRQ* are both logic 0s, along with A15–A11 being 01101. For another 2716 in the same system, these address signals might be in the state 00000 or 11010. Typically, the resulting logic 0 signal is wired to the $\overline{CE}$ or $\overline{CS}$ input of the memory chip. Remember, the bar over a signal name, like the asterisk, indicates that the signal (pin) is active in the logic 0 state. So, to enable the 2716 and have it place information on its data outputs, the $\overline{CE}$ input would have to be a logic 0. To enable the chip and the $\overline{OE}$ (output enable) pin, the input would have to be a logic 0 so that the data is present on the EPROM's outputs.

In the case of R/W memory, the MEMRQ* signal would be gated with some combination of address signals and the result would be used to enable the R/W memory chip. The WR* signal would then be wired to the chip and would be used to indicate whether a read or write operation is taking place. If the microprocessor is communicating with memory and WR* is a logic 1, then a *memory-read* operation is taking place (by default). If the microprocessor is communicating with memory and WR* is a logic 0, then a *memory-write* operation must be taking place.

MEMORY MAPS

Not only are you probably interested in interfacing, but at some point, your STD bus microcomputer is going to have to be programmed, using either assembly language, BASIC, FORTH, or some other language. In this book, we will be primarily interested in interfacing, and we will use assembly language to control the interfaces. However, before any programming is done, it is imperative that you know how much memory is in your system and how much you can use for your programs and data. Remember, programs and some data will probably be stored in EPROM, and data acquired from an experiment or control process will probably be stored in R/W memory.

In general, all of the processors currently available for the STD bus can be grouped into two classes—the 8085 types (8085, Z-80, NSC800, and 8088) and the 6800 types (6800, 6809, and 6502). As we shall see, this classification applies not only to how these processors use memory but, also, to how they communicate with peripherals.

When all of these processors were designed, the designers had to make some decisions about how memory would be used. Thus, we cannot store just anything we want (data or instructions) in just any of the 65,536 possible memory locations. The chip designers have specified that specific information would be stored in a few memory locations in order for the processsors to work.

When these chips were designed, the memory location that contains the first instruction, or the location that points to the first instruction that is executed, was fixed. For 8085-type processors, the first instruction that is executed must be contained in memory location Zero (0000_{16}). For the 6800-type processors, the last few memory locations (up near $FFFC_{16}$) contain the *address* of the memory location that contains the first instruction that is to be executed. Thus, 8085-type processors are reset to "lo" memory and 6800-type processors are reset to "hi" memory. These sections of memory are usually EPROM; so when the microprocessors are reset, they start executing a program. It is possible to have R/W memory in these sections of the microprocessor's "memory map" but we won't be concerned with how this is done.

Two other sections of memory that you must be concerned with are: (1) where the stack is located, and (2) where the interrupt vectors are located. The sections of memory that are used by these functions are summarized in Table 1-5.

Table 1-5. Fixed-Function Memory Locations for STD Bus Processors

CPU	Reset To	Interrupted To	Stack
8085	0000*	0000–003C	anywhere
Z-80	0000*	0000–0038, anywhere	anywhere
NSC800	0000*	0000–0038, anywhere	anywhere
8088	FFFF0†	00000–003FF	anywhere
6800	FFFE†	FFF8–FFFD	anywhere
6809(E)	FFFE†	FFF0–FFFD	anywhere
6502	FFFC†	FFFE, FFFA	0100–01FF

* Contains the first instruction to be executed.
† Contains the address of the first instruction to be executed.

I/O DEVICES

What is an input/output device? Very simple, an I/O device is an electrical device that the microcomputer communicates with, other than memory. Thus, a 2716 EPROM is probably not an I/O device (although it could be wired as an I/O device), and a paper-tape reader is not memory. Typically, I/O devices can be separated into those devices that *transfer* data and those devices that are just *controlled* by the microcomputer. There are also a large number of devices that *transfer* data *and are controlled* by the microcomputer.

As an example, information is usually just output to a simple LED display. The computer doesn't have any control over the display. It can't turn three digits on and four digits off, and the LED display doesn't send a signal to the microcomputer saying that it can accept data, etc. (Of course, there are some very complex LED displays that can do this.) Thus, for a simple display, information is just output to the display.

A simple stepping motor is an example of a control-only device. With this type of peripheral, the microcomputer may generate a sequence of pulses that are used to control the position of the stepping motor's shaft. No *information* is transferred between the microcomputer and the motor. The computer would not output a value of 10111000 to the motor and, at some later time, read back a value of 00111010. Again, in some complex stepping motors, the microcomputer might be able to transfer data.

A large number of peripherals transfer information between and are controlled by the microcomputer. In a floppy disk interface, data are read from and written out to the floppy disk. The microcomputer also generates pulses that are used to step the head in and out and, also, "drop" the head onto the disk. In an asynchronous serial communications interface, which is used by the microcomputer to communicate with a teletypewriter or a CRT, information is transferred between the two devices, and pulses are used to initiate the transmission of data. Some devices can be grouped as data-only, control-only, or data/control devices. Some typical examples are the following data-only peripherals:

  Digital-to-analog converters
  LED displays
  Thumbwheel switches
  pH meters
  Current sensors
  Thermocouples
  Light sensors

Then, there are the control-only peripherals:
  Stepping motors
  Digital waveform generators/sequencers
  Solid-state relays
  Limit switches

Finally, we have the data/control peripherals:
  Floppy disks
  Paper-tape readers
  Analog-to-digital converters
  Keyboards
  Printers
  EPROM programmers
  Asynchronous communications interfaces
  Microwave ovens
  Traffic light controllers
  Home security systems
  Automatic bank tellers
  Heating/cooling systems
  Digital plotters

Just as the microprocessors are classified as to their "dedicated" memory locations, they can also be classified the same way when discussing their I/O capabilities. The 8085-type processors can communicate with I/O devices using two different types of interfaces and assembly language instructions, while the 6800-type processors can only use one type. This means that the 8085, Z-80, NSC800, and the 8088 microprocessors can communicate with peripherals using either *accumulator I/O* or *memory-mapped I/O*. The 6800, 6809 and 6502 processors are limited to just *memory-mapped I/O*.

MEMORY-MAPPED I/O

Memory-mapped I/O is an interfacing technique where the I/O device or peripheral looks like one or more *memory locations* to the microprocessor. In order to "look like" a memory location, the peripheral is "identified" by a 16-bit memory address on the address bus and uses the data bus, along with the *RD*, WR*, and MEMRQ* control signals*. The important point is that the microprocessor "thinks" that it is communicating with one or more memory locations when it is really communicating with an I/O device.

This means that if a peripheral has an address of $BC00_{16}$, there must be no memory (read-only or read/write) selected when this address is present on the address bus. Thus, normally, it would not be possible to have 65,536 words of memory, along with 10–15 memory-mapped I/O devices. In some cases, it is possible to have 65K of memory along with a number of memory-mapped peripherals, but only if the MEMEX (memory expansion) signal is used. By using this signal, it is possible to have the microprocessor select one of many possible 65K "pages" (where a page might consist of just I/O devices, just memory, or a combination of both). If the MEMEX signal is wired to ground on your CPU card, you are limited to just one "page" of 65K and, within this page, you will have to have all memory-mapped I/O devices and all memory. Thus, you might have 64,508 words of memory, and a reserve of 1024 memory addresses for memory-mapped I/O devices.

ACCUMULATOR I/O

If accumulator I/O is used, an 8-bit address is generated by the microprocessor, and is present on A0–A7. At the same time, control signals RD* or WR* (along with IORQ*) are generated. As you can see, the two differences between memory-mapped I/O and accumulator I/O are the number of address bits used and the control signals generated (either MEMRQ* or IORQ*). Since the MEMRQ* signal is not used in an accumulator I/O interface, the microprocessor *can* have 65,536 memory locations, along with 15–20 interfaces. Thus, none of the microprocessor's "memory address space" is used up by the accumulator I/O peripherals.

However, one disadvantage of accumulator I/O is the fact that there can only be up to 256 8-bit input devices and 256 8-bit output devices. However, in most situations, you won't come close to having this many devices interfaced to your microcomputer.

As you will see, there are two different types of assembly language instructions that can be used to communicate with peripherals. Any peripheral that is interfaced using memory-mapped I/O techniques can be accessed using the same instructions that the microprocessor uses to transfer information between itself and memory (memory reference instructions). To communicate with accumulator I/O devices, special accumulator I/O instructions must be executed.

CPU COMPATIBILITY

At this point, something may have struck you as being a little odd. If the 6800-type microprocessors are only capable of memory-mapped I/O, how can CPU cards based on these chips generate the accumulator I/O signals (IORQ* and an 8-bit address)? Remember, they can only perform memory-mapped I/O.

These signals are indeed generated on 6800-type CPU cards, and they are generated in a very clever way. In most cases, the CPU card designer has selected a block of 256 memory addresses which, when generated by the CPU chip, cause the IORQ* signal to be generated. This takes anywhere from one to three ICs (usually decoders). Therefore, in a 6800 system, for example, if any memory address is generated where the eight most-significant bits (MSBs) of the address bus are 00111100 or 3CH, the IORQ* signal will be generated. At the same time, this logic prevents the MEMRQ* signal, which is generated whenever the CPU accesses memory, from being sent to the edge connector of the CPU card.

Let's assume that we want the 6800 to communicate with accumulator I/O device 45H. How would this be done? In order to do this, we would have to program the 6800 to communicate with "memory location" 3C45H. When this address is present on the address pins of the CPU chip, external logic recognizes that the 8 MSBs of the address bits contain 3CH. This causes the IORQ* signal to be generated, and prevents the MEMRQ* signal from going out to the edge connector. Of course, as long as "memory" is being addressed, the RD* or WR* signals will be generated.

At the same time, the complete 16-bit address is present on the address bus. However, in accumulator I/O, the peripheral only monitors the 8 LSBs on the address bus, so with a "memory address" of 3C45H, only the 45H in this address is monitored by the accumulator I/O device.

On some 6800-type CPU cards, the block of 256 memory addresses that cause the IORQ* signal to be generated will be fixed, while on other cards, the user can change this address via a group of 8 DIP switches. At this time, no standard has been established as to which of the 256 possible blocks of addresses will generate the IORQ* signal. On some systems, it will be C000H–C0FFH, FE00–FEFFH, etc.

I/O INTERFACE COMPATIBILITY

In order to have an interface that will work with *any* CPU card, only two requirements have to be met: (1) it must be capable of memory-mapped I/O, and (2) it must not use any MOS/LSI chips. The reason that the card must be at least capable of memory-mapped I/O is that some 6800-type CPU cards may be designed that can't generate the IORQ* signal. Most can, but you never know . . . . Of course, many boards are capable of both memory-mapped I/O and accumulator I/O. Thus, by changing the setting of a few switches or jumpers, you can switch from one mode of operation to the other.

How do you suppose that this is done? Really, it is relatively easy. In memory-mapped I/O, the MEMRQ* signal has to be used, along a 16-bit address. In accumulator I/O, IORQ* is used along with an 8-bit address. Of course, in either technique, RD* and/or WR* is used. Thus, in order to design a board that can be used with either technique, all the designer has to do is to provide an area on the board where either MEMRQ* or IORQ* can be jumpered or switched into the control logic. To monitor an address on the address bus, the logic for monitoring all 16 address bits must be present on the board. However, the user can select, probably by jumpers, whether all 16 bits of the address bus will be used to monitor an address, or whether just the 8 LSBs of the address will be monitored. As you will see in Chapter 2, this is all pretty easy to do.

DATA TRANSFER TIMING

The basic reason why some interface cards won't work with some CPU cards is *timing*. As you have seen, the IORQ* signal can be generated by some non8085-type processors, so if the interface cards are designed properly, they should be able to be used with any processor. However, problems do occur when a processor tries to transfer data to a peripheral chip at a speed that is faster than the peripheral chip was designed to send or receive data.

As an example, let's examine the timing of a typical Z-80 CPU card. The Read timing waveforms are shown in Fig. 1-3 and the Write timing waveforms for the CPU card are shown in Fig. 1-4. As you can see, all operations that the Z-80 performs are referenced to the clock (CLOCK) which, in a typical system, may be either 4 MHz or 2.5 MHz. Regardless of the type of memory operation being performed, the Z-80 always places an address on the address bus and generates

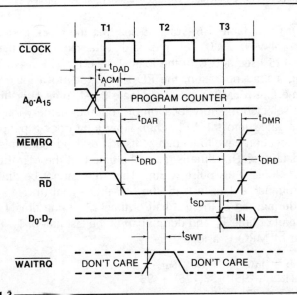

FIGURE 1-3
Typical timing for a Z-80 memory-read operation. *(Courtesy Mostek Corp.)*

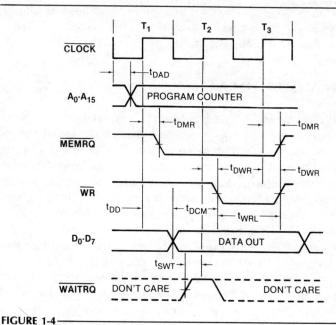

FIGURE 1-4

Typical timing for a Z-80 memory-write operation. *(Courtesy Mostek Corp.)*

MEMRQ. From Table 1-6, you can see that the clock period is 250 nsec for a 4-MHz Z-80. This means that the clock is in the logic 0 state for 125 nsec and is in the logic 1 state for 125 nsec.

During a read operation, the $\overline{RD}$ signal goes to a logic 0 at the same time that $\overline{MEMRQ}$ goes to a logic 0. At some later time, data from the selected memory location or from the memory-mapped I/O device is gated into the CPU chip. How long does a peripheral or memory location have to recognize its address and place information on the data bus? This time is really the time that the $\overline{MEMRQ}$ or $\overline{RD}$ signal is a logic 0 (its pulse width). This time cannot be directly obtained from Table 1-6. Instead, the $\overline{MEMRQ}$ signal (in Fig. 1-3) is a logic 0 during part of the last half of time T_1, during time T_2, during the first part of time T_3, and during part of the last half of T_3. Thus, the time that $\overline{MEMRQ}$ is a logic 0 is:

$$T = t_{\frac{1}{2}} - t_{DAR} + 3(t_{\frac{1}{2}}) + t_{DMR}$$
$$= (125 - 95) + 3(125) + 95$$
$$= 30 + 375 + 95$$
$$= 500 \text{ nsec}$$

where,

$t_{\frac{1}{2}}$ = 125 nsec (4-MHz clock).

Note, however, that the data must be stable 80 nsec (t_{SD}) prior to the positive edge of the clock during time T_3. This reduces the time from 500 nsec to 500 nsec − t_{DMR} − 80 nsec, or 325 nsec. Thus, the memory and peripherals must have *access times* of 325 nsec or less.

If you go through the same type of calculations for a write operation, you will find that memory and memory-mapped I/O devices must have access times of 305 nsec or less. Note that this is 20 nsec less than the access time required for a read operation, because t_{DWR} is 20 nsec less than t_{DMR} (Table 1-6). Now that you have seen the basic timing for the Z-80 CPU card, let's see why the different families of MOS/LSI chips cannot be used interchangeably.

CHIP INCOMPATIBILITY

The reason that problems arise when using Z-80 peripheral chips with a 6800 device, or 6502 peripheral chips with an 8085 processor, etc., is due to the fact that the MOS/LSI chips are relatively slow. A typical TTL chip can respond to a change in a signal level in only 10–20 nsec, while an MOS/LSI chip may require 250–400 nsec to sense the change. As an example, let's look at an interface between the Z-80 and an MC6852 serial communications chip. The timing for a read operation is shown in Fig. 1-5 and the values for these times are summarized in Table 1-7.

Let's assume that the "B" version of the chip is being used. It is the fastest of the three (and the most expensive). From Fig. 1-5, you can see that the cycle time of the Enable input must be at least 500 nsec. If this is wired to the 250-nsec clock of a 4-MHz Z-80, the 6852 won't work. Likewise, a 2.5-MHz Z-80 (400-nsec cycle time) is also too fast for this type of device. Even if you slow the CPU clock down, the chip still won't work with a Z-80 because the read operation for the 6852 chip requires just 1 clock cycle, while a typical Z-80 operation requires 3 clock cycles. The point is, you would need some pretty fancy clocked logic on either the CPU card or on the peripheral card in order to use these two devices in the same system. Of course, the CPU card manufacturer doesn't want to have to do this and neither does the peripheral card manufacturer. Thus, you, basically, cannot use chips that are designed for use with one CPU on a

35

Table 1-6. Timing Values for a 4-MHz Z-80

Bus Signal	Symbol	Description	Min	Max	Units
/CLOCK	t_c	Clock Period.		250	nS
A_0-A_{15}	t_{DAD}	Address output delay.		130	nS
	t_{ACM}	Address stable prior to /MEMRQ.	75		nS
	t_{ACI}	Address stable prior to /IORQ, /RD, /WR (I/O cycle).	180		nS
D_0-D_7	t_{SDO}	Data setup time during /M1 cycle.	70		nS
	t_{SDR}	Data setup time during memory-read cycle, I/O-read cycle, or interrupt-acknowledge cycle.	80		nS
	t_{DCM}	Data stable prior to /WR (memory write).	80		nS
	t_{DD}	Data output delay.	155		nS
	t_{DCI}	Data stable prior to /WR (I/O write).	−30		nS
	t_{CDF}	Data stable from /WR.	70		nS
/MEMRQ	t_{DMR}	/MEMRQ delay from /CLOCK.		95	nS
	t_{MRH}	/MEMRQ pulse width high.	105		nS
	t_{MRL}	/MEMRQ pulse width low.	220		nS
/IORQ	t_{DIR}	/IORQ delay from /CLOCK.		85	nS
/RD	t_{DRD}	/RD delay from /CLOCK.		95	nS

Table 1-6. Cont. Timing Values for a 4-MHz Z-80

Bus Signal	Symbol	Description	Min	Max	Units
/WR	t_{DWR}	/WR delay from /CLOCK.		75	nS
/M1	t_{DMI}	/M1 delay from /CLOCK.		110	nS
/REFRESH	t_{DRF}	/REFRESH delay from /CLOCK.		130	nS
/WAITRQ	t_{SWT}	/WAITRQ setup time.	85		nS
/INTRQ	t_{SINT}	/INTRQ setup time.	95		nS
/NMIRQ	t_{SNMI}	/NMIRQ setup time.	85		nS
/BUSRQ	t_{SBR}	/BUSRQ setup time.	55		nS
/BUSAK	t_{DBA}	/BUSAK delay from /CLOCK.		110	nS
	t_{TS}	Tri-state buffer delay from /BUSAK for A_0-A_{15}, D_0-D_7, /MEMRQ, IORQ, /M1, /RD, /WR, /REFRESH, and /INTAK.		25	nS

Courtesy Mostek Corp.

different CPU. The possible exceptions to this are some of the 8085-type peripheral chips that can be used with some of the 8085-type CPUs (a Z-80 peripheral with the 8085, an 8085 peripheral with the 8088, etc.), and some of the 6800-type peripherals can be used with some of the 6800-type CPUs. In general, though, you don't want to mix families, because of these timing problems.

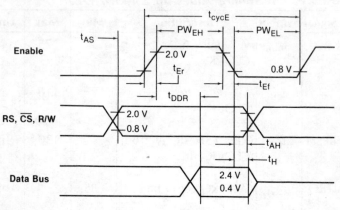

FIGURE 1-5
Timing waveforms for an MC6852 communications chip. *(Courtesy Motorola Semiconductor Products, Inc.)*

Note that the inability to use the 6852 with the Z-80 is not due to any deficiencies in either the 6852 or in the Z-80. They were just designed by different groups of engineers who approached the design

Table 1-7. Timing Values for an MC6852 Read Operation

Characteristic	Symbol	MC6852 Min	MC6852 Max	MC68A52 Min	MC68A52 Max	MC68B52 Min	MC68B52 Max	Unit
Enable Cycle Time	t_{cycE}	1.0	—	0.666	—	0.5	—	μsec
Enable Pulse Width, High	PW_{EH}	0.45	25	0.28	25	0.22	25	μsec
Enable Pulse Width, Low	PW_{EL}	0.43	—	0.28	—	0.21	—	μsec
Setup Time, Address and R/$\overline{W}$ Valid to Enable Positive Transition	t_{AS}	160	—	140	—	70	—	nsec
Data Delay Time	t_{DDR}	—	320	—	220	—	180	nsec
Data Hold Time	t_H	10	—	10	—	10	—	nsec
Address Hold Time	t_{AH}	10	—	10	—	10	—	nsec
Rise and Fall Time for Enable Input	t_{Er}, t_{Ef}	—	25	—	25	—	25	nsec

Courtesy Motorola Semiconductor

problem of the chips in different ways. It would be very difficult, if not impossible, to design a CPU chip that works with every possible peripheral chip, and it would be just as difficult to design a peripheral chip that could be used with every possible CPU chip.

NONSTANDARD SIGNALS

At the beginning of this chapter, we mentioned that four of the STD bus signals no longer had one meaning or function, but were processor specific (Table 1-3). Thus, when purchasing cards, you should find out from the manufacturer if these signals are used by any of the cards in which you are interested. Remember, these signals are generated by the CPU card and may be used by memory or peripheral cards. If a card uses one of these "special" signals, then it is probably only STD-6800, STD-6502, etc., compatible.

CONCLUSION

In this chapter, we have examined the STD bus standard—the mechanical, physical, and electrical standards. The types of CPU chips (cards) that are available were also discussed, along with the signals that these chips and cards generate.

As you have seen, an STD bus microcomputer consists of not only a CPU card but, also, of a number of memory and peripheral cards. The memory cards may contain EPROM and/or read/write memory (which can be either dynamic or static). Peripheral cards can be used to output information to lights, a floppy disk, or to 7-segment displays, and they can also be used to input information from an ASCII keyboard, from thumbwheel switches, or a paper-tape reader.

Regardless of the type of peripheral card used, it will consist of either input ports, output ports, or both. These ports may recognize a 16-bit memory address (memory-mapped I/O) or an 8-bit address (accumulator I/O). These cards will also be connected to the data bus, so that information can be transferred between the peripheral card and the CPU, and they will use certain combinations of control signals so that data are only put on the data bus, or taken off of the data bus, at very specific times.

I/O Device Addressing
Chapter 2

In Chapter 1, you saw that the CPU cards generated the signals RD*, WR*, IORQ*, and MEMRQ*. In addition, they could also generate some special or processor-specific signals, which were present on pins 37 through 40. In most interfaces, standard control signals are used, while in interfaces that use MOS/LSI chips, processor-specific signals are used. In this chapter, you will see how peripherals or interfaces are addressed. This means that we will be using the 8- or 16-bit address that the processor generates in order to identify one interface among many. In general, this circuitry will generate a short logic 1 or logic 0 pulse when the peripheral's address or "name" is present on 8 or 16 lines of the address bus. This pulse, when used with signals RD*, WR*, and either IORQ* or MEMRQ*, can be used to transfer information or control a peripheral.

As you will see, there are many different circuits that can be used to provide this decoding function. We won't be able to show you every last combination or permutation of circuitry, but we will discuss a good representative sample. Starting in this chapter, we will assume that you have some background in digital logic; either TTL or CMOS will be just fine.

ADDRESS AND CONTROL SIGNALS

It is important to remember that there are two different techniques that can be used to communicate with peripheral devices. You can use memory-mapped I/O or "accumulator" I/O. The use of the STD bus with these two different techniques is summarized in Table 2-1. By reviewing this table, you can see that in memory-mapped I/O, a 16-bit address has to be decoded, while in accumulator I/O, only an 8-bit address has to be decoded. Likewise, the MEMRQ* signal indicates to the user that a memory or memory-mapped I/O operation is taking place and the IORQ* signal indicates that an accumulator I/O operation is taking place.

As you will see, there are also differences in the assembly language instructions that can be used to access these different types of interfaces. In general, to keep the software as simple as possible, the same basic program will be used for the 8085 and the Z-80. Completely different programs will be used for the 6800 and the 6502.

DEVICE ADDRESSING

Each I/O device that is used in an STD bus computer must be able to recognize its own device address. In the case of memory-mapped I/O, the peripheral must monitor A0–A15 for the occurrence of its address. For accumulator I/O, the peripheral must monitor A0–A7 for its address.

There are three basic schemes that can be used by an I/O device to monitor the address bus for a specific address. These are:

1. Gating—The detection of a specific combination of logic signals.
2. Decoding—A flexible gating scheme where many addresses can be detected.
3. Comparing—The comparing of a preset or known address with the address bus signals received until the two are equal.

The combinations for these three techniques are endless. Examples of each of these three basic decoding schemes will be described in the following pages.

USING GATES FOR ADDRESS DECODING

In a scheme for decoding device addresses, where individual gates are used, the address must be known so that the gates can be properly wired to the address bus. In this example, we will use the accumulator I/O device address of 01111011_2, or 7BH. Since NAND and

Table 2-1. A Comparison of Accumulator I/O and Memory-Mapped I/O

	Max. I/O Devices	Address Bus Use
Accumulator I/O	256	A0–A7
Memory-mapped I/O	65,536	A0–A15

	Control Signal Use	Data Bus Use
Accumulator I/O	RD*, WR*, IORQ*	D0–D7
Memory-mapped I/O	RD*, WR*, MEMRQ*	D0–D7

AND gates are the predominant types of gating logic available, we will use these types of gates in the decoder design. To refresh your memory, the pin configurations for several types of NAND and AND gates are shown in Fig. 2-1. Since inverters such as the SN7404 are often found in device address circuits, a pin configuration for this type of chip has also been included in Fig. 2-1. Generalized truth tables for a two-input AND gate, a two-input NAND gate, and an inverter are shown in Table 2-2. *In all cases, the logic 1 state is the higher voltage (+2.8 to +5 volts) while the logic 0 state is the lower voltage (0.0 to 0.8 volt).* The NAND gate devices are available with 2, 3, 4, 8, and 13 inputs, while AND gates devices are available that have 2, 3, and 4 inputs.

Since the unique output state (logic 1 for the AND gate and logic 0 for the NAND gate) occurs only when *all* of the inputs to an AND or NAND gate are all logic ones, we will have to configure the accumulator I/O address 7BH (01111011) so that it generates eight logic 1s at the inputs to an 8-input NAND gate when this address is present on A0–A7 of the address bus. This simply means that the logic 0s at positions A7 and A2 must be inverted, as shown in Fig. 2-2.

The output of an SN74LS30 NAND gate will be a logic zero *only* when *all of the inputs are logic one*. Thus, address 7BH (01111011) will cause the NAND gate output to go to a logic zero (Fig. 2-2). If a logic one is required by the I/O device when this address is present, it can be obtained by simply inverting the NAND gate output, as shown. The inverters are SN74LS04 devices. The outputs are labeled DEVICE ADDRESS and DEVICE ADDRESS, respectively, to indicate the logic state that Is unique, or to indicate the state that will be used to cause an action to take place at the I/O device.

Remember that the bar over a signal name, like the asterisk, indicates that the signal is active in the logic 0 state. Most STD bus manufacturers use one or the other of these "nomenclatures."

It is also possible to decode a memory address for a memory-mapped I/O device by using the preceding technique (Fig. 2-3). While these gating schemes are effective in decoding a single address, and are relatively inexpensive, this technique is very inflexible. A more flexible scheme is shown in Fig. 2-4. This circuit illustrates the use of a gating scheme where inverters may be used to invert individual address bits. As shown, jumpers are required to connect one part of the circuit to another. The bits may also be used without any inversion. The eight jumpers allow the device address to be preset, as illustrated in Fig. 2-5, for the address 96H (10010110_2). If the programmable circuit in Fig. 2-4 is used to "detect" or decode the device address, any one of 256 possible addresses may be preset. However, only one address may be preset at a time. The same flexible scheme can also be used with the memory-mapped I/O, where sixteen inverters, two 8-input NAND gates and one OR gate would be required. The OR gate can be used to gate both the NAND gate outputs together.

The programmable gating circuit provides a lot of flexibility but it can detect only *one* of 256 (or 65,536) possible addresses. This is a limitation, particularly when a number of I/O devices are located on the same STD bus card; each will require its own gating circuit. You will see how this limitation can be overcome in a later section of this chapter.

Unfortunately, the gating schemes that you have been shown will not uniquely identify the address of an *I/O device*. The address bus is also used to address memory locations so there is an excellent

Table 2-2. Truth Tables for a Two-Input AND Gate, a Two-Input NAND Gate, and an Inverter

AND Gate		NAND Gate		Inverter	
Inputs A B	Output Q	Inputs A B	Output Q	Input A	Output Q
0 0	0	0 0	1	0	1
0 1	0	0 1	1	1	0
1 0	0	1 0	1		
1 1	1	1 1	0		

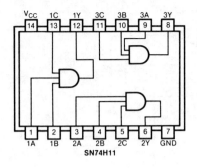

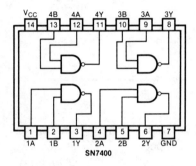

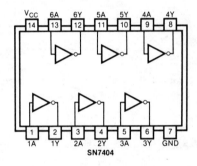

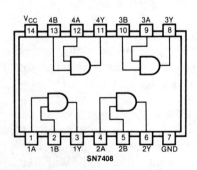

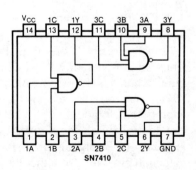

Figure 2-1
Pin configurations for various AND and NAND gates and an inverter.

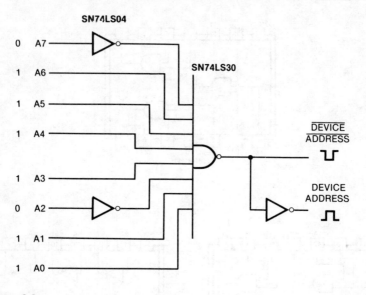

Figure 2-2

Using inverters and a NAND gate to detect address 7BH.

chance that the 8-bit address of the I/O device will be present on the address bus at various times. It will not appear as an I/O device address, *but as the 8 least-significant address bits in a 16-bit memory address*. Thus, the following memory addresses 00111000 01111011, 01011101 01111011, 00000001 01111011, and 10101001 01111011 would *all* activate the device address-detecting circuit shown in Fig. 2-2. Obviously, there must be some additional circuitry that will allow us to differentiate whether an I/O address or a memory address is present on the address bus.

As you already know, the CPU card generates an IORQ* signal when an I/O operation is taking place, and it generates an MEMRQ* signal when a memory operation is taking place (Table 2-1). Thus, *the IORQ* signal is used* to indicate that an I/O address is present on the address bus. In addition, *the RD* and WR* signals are used* to indicate to peripherals the direction of the data flow on the data bus. In Fig. 2-6, these signals are gated with the output of the address-detection circuitry. Note that the IORQ* signal is first gated with RD* and WR*, to generate the IORD* and IOWR* signals. These signals will only be generated when the microprocessor chip is executing an

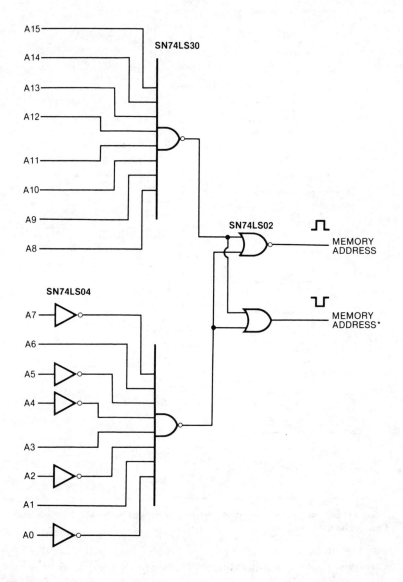

Figure 2-3
Using gates to detect a 16-bit memory address.

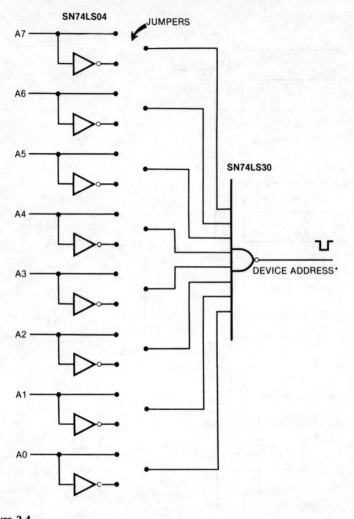

Figure 2-4

Using jumpers to connect (program) the device address into the address-detection logic section.

input or an output instruction. Since some CPUs (6502, 6800, 6809) can only address memory, these signals will be generated whenever the microprocessor performs a read or write operation *and* an address, within a specific block of 256 memory addresses, is present on the address bus.

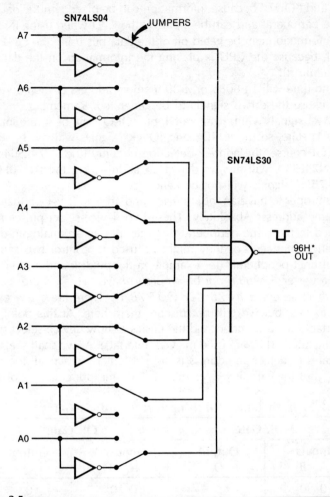

Figure 2-5

Adding jumpers to detect I/O address 96H (10010110).

By way of review, the truth tables for an OR gate (SN7432) and a NOR gate (SN7402) are given in Table 2-3. The pin configurations for these two chips are shown in Fig. 2-7.

As was shown in Fig. 2-6, the IORD* and IOWR* signals are gated with the output of the address-detection circuitry. The two logic 0 signals are gated with an OR gate to form the OUT 7BH* signal, and a NOR gate is used to form the IN 7BH signal. These signals are called

49

"IN" and "OUT" because information can be placed on the data bus by the peripheral and can be *input* by the CPU when using IN 7BH, and information can be gated off of the data bus using the OUT 7BH signal, because the CPU is placing the information on the data bus (*outputting* it).

At no time will an 8085 or Z-80 memory address or memory operation cause these two signals to be generated. Remember, the RD* and WR* signals will be generated, but IORQ* will be in the inactive (logic 1) state, so the IORD* and IOWR* signals will not be generated. Of course, the address-detection circuitry (the SN74LS04s and the SN74LS30) will always detect an address, but the IN 7BH and OUT 7BH* signals will not be generated.

Is it proper to have an *input device* and an *output device* that have the same address? Absolutely. These two device-select pulses might be used in the same peripheral (for example, in an analog-to-digital converter interface), or they might be used to control two separate and distinct peripherals (for example, a floppy disk and an X-Y plotter). *However, it would not be proper to have two (or more) peripherals that use either IN 7BH or OUT 7BH**. There are a few exceptions to this, but we will not discuss them here. At this point, it is important that you understand the basics of how device-select pulses are generated. They are used in *every* interface. As we shall see, using just gates to detect an address is not particularly practical (too many chips), but this circuitry does demonstrate the principles involved.

Table 2-3. Truth Tables for 2-Input NOR and OR Gates

NOR Gate		OR Gate	
Inputs A B	Output Q	Inputs A B	Output Q
0 0	1	0 0	0
0 1	0	0 1	1
1 0	0	1 0	1
1 1	0	1 1	1

USING DECODERS

In many cases, it is easier to use *decoder circuits* in place of the circuitry that is based on individual gates that you have seen previously. Why are decoders so useful? Perhaps we should take a look at

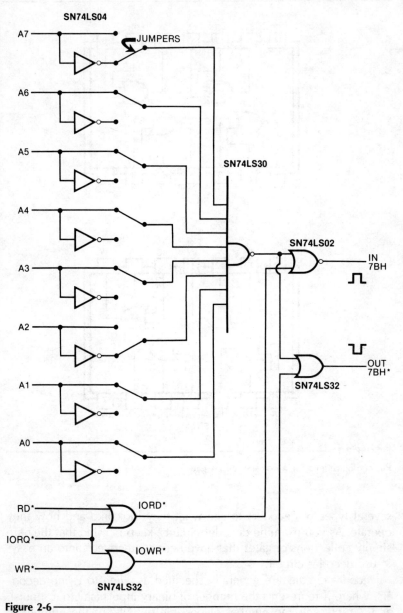

Figure 2-6

Using IORD* and IOWR* to generate device select pulses for I/O device synchronization.

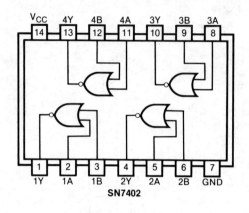

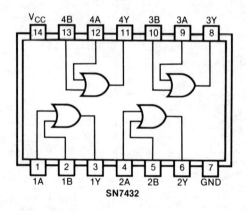

FIGURE 2-7

Pin configurations for a NOR and an OR gate.

several types of decoders to see what they look like and how they operate. As you examine decoder circuits, keep in mind that they are simply collections of gates that have been "integrated" into an easy-to-use decoder circuit.

Decoder circuits are generally specified as *x-line to y-line decoders*, where X represents the number of binary inputs (say four inputs), and Y represents the number of possible outputs, or the number of different binary states present on the X inputs. Thus, for four inputs, there would be sixteen possible outputs, creating a 4-line to 16-line decoder. This is, in fact, a real circuit (the SN74LS154), as you will see later.

Each of the binary inputs has two possible states, a logic 1 and a logic 0. These inputs are all independent of one another. The outputs are also binary, in the sense that they have two possible states, but they are *not independent*. There will be only *one* unique output from the decoder, representing the value or "weight" that is present on the binary inputs. In most cases, the unique output state is a logic 0, with the other outputs in their logic 1 state.

A typical decoder integrated circuit is the SN74LS139. This integrated circuit actually contains two independent 2- to 4-line decoders, as shown in Fig. 2-8. The truth table for the SN74LS139 is shown in Table 2-4.

Of course, the truth table applies to both of the decoders within the SN74LS139 dual integrated-circuit chip. Most decoder circuits contain an *enable input,* so that the decoder can be turned on and off by

Table 2-4. Truth Table for the SN74LS139 2- to 4-Line Decoder

Inputs			Outputs			
Enable	Select					
G	B	A	Y0	Y1	Y2	Y3
H	X	X	H	H	H	H
L	L	L	L	H	H	H
L	L	H	H	L	H	H
L	H	L	H	H	L	H
L	H	H	H	H	H	L

H = high level; L = low level; X = irrelevant (don't care)

either a logic 1 or logic 0 input. This is the function of the ENABLE or "G" input of each of the decoders in the SN74LS139. Note that when the "G" input is a logic 1, all of the decoder outputs are forced into the logic 1 state, regardless of the state of the A and B inputs. When the "G" input is a logic 0, the decoder will function, so the active output will reflect the combination of 1s and 0s on the inputs. The "G" input allows the decoder to be gated on and off. In the OFF state, the power is not removed, but the outputs are all forced to the logic 1 state.

Now, let's examine a simple, and rather trivial, example of a device-address decoder based on a 2- to 4-line decoder. We will assume that there are only a few I/O devices so that the decoders in the

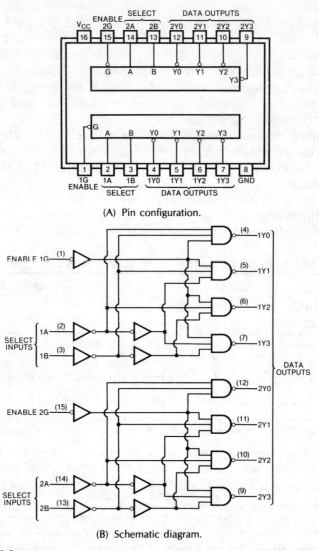

(A) Pin configuration.

(B) Schematic diagram.

FIGURE 2-8

The SN74LS139 decoder chip.

SN74LS139 will be sufficient. In this simple circuit (Fig. 2-9), only two address bits have been decoded, the rest have been ignored. Note the use of the IORQ* signal, which is active in the logic 0 state,

54

to enable the decoder. Thus, the decoder is only enabled when the CPU card is communicating with an accumulator I/O device. The NOR and OR gates are used to gate the decoded addresses with the RD* and WR* signals. The signals generated by these gates are the device-select pulses.

The device-select pulses have been noted as IN X, IN Y*, and OUT Y, since there is not a single *specific* address that will cause each device-select pulse to be generated. Device addresses 00000010, 11110110, and 11111010 will all cause the IN X device-select pulses to be generated during an I/O read operation. This *nonabsolute* device addressing results because address bits A2–A7 have not been used in the address decoder design. Nonabsolute addressing means that there are several addresses that will actuate the selected device. The circuit shown in Fig. 2-9 will decode four addresses and, thus,

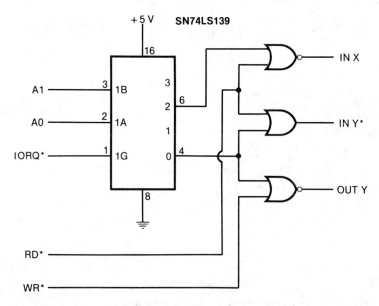

Figure 2-9
Using an SN74LS139 chip as an I/O device address decoder.

eight individual devices can be selected (four input devices and four output devices). Additional NOR and OR gates are required, though. In a small system, this may be adequate, although the decoding scheme does not provide a great deal of flexibility in allowing the addition of new I/O devices beyond the eight original ones. Even though this design is not very flexible, let's take a closer look at it since it will

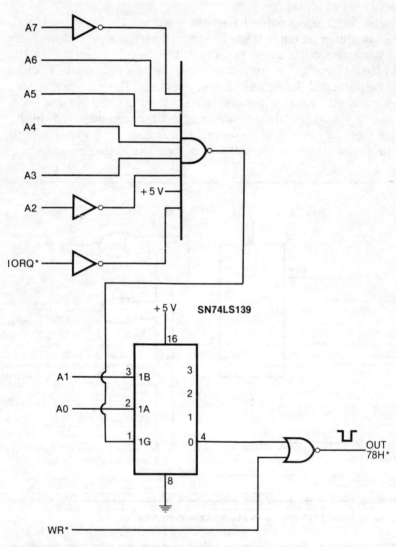

FIGURE 2-10

Absolutely decoding of an I/O address using gates, inverters, and an SN74LS139 decoder.

allow us to develop two other concepts that can be applied to other decoder schemes.

In Fig. 2-9, the enable input (G) of the decoder is wired to IORQ*. This input can also be used so that the 8-bit address is absolutely decoded. A 8-input NAND gate circuit can be used to supply an enabling signal to the decoder only when a preset pattern of address bits is present on A2–A7. You have already seen the use of an 8-input NAND gate in a device-address decoding scheme. The circuit shown in Fig. 2-10 is simply a combination of the two.

In the design shown in Fig. 2-10, the decoder is only enabled when the A7–A2 address lines contain 011110 and when the IORQ* signal is generated. Once the decoder is enabled, it decodes A1 and A0. In this circuit, then, the decoder outputs will correspond to addresses 78H, 79H, 7AH, and 7BH. Only the OUT 78H* device-select signal

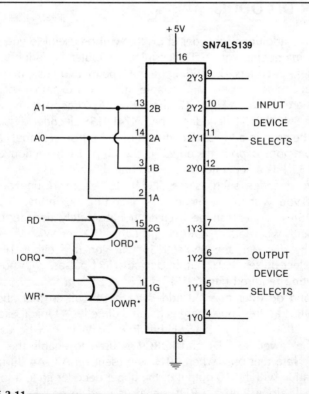

FIGURE 2-11
Using IORD* and IOWR* to enable the dual decoders contained in an SN74LS139.

has been generated in the example that is shown. Again, an OR or NOR gate will be required for each device-select signal required.

An alternate approach is to use both the decoders in the SN74LS139 chip and use IORD* and IOWR* to enable the decoders. In this way, the address selection is again *nonabsolute,* but the device-select pulse gating is performed within the chip. This is shown in Fig. 2-11. The NOR and OR gates are no longer required in order for each device-select pulse to be generated.

While this circuit may not be particularly useful, it illustrates the use of the enable input of the decoder in generating the device-select pulses. The decoder gating or enabling input may be used for device-select pulse generation or for absolute address decoding. In some cases, it may be used for both.

LARGER DECODERS

There are additional decoder chips that will be useful to you when you start interfacing your STD bus microcomputer to peripheral devices. These decoders, depending on the type that you use, may have additional inputs, outputs, and enable lines. Two examples of these decoders are shown in Fig. 2-12 (the SN74LS138 decoder) and in Fig. 2-13 (the SN74LS154 decoder). The SN74LS155 decoder (Fig. 2-14) can also be used in address decoding designs, but select inputs A and B are common to *both* decoders in the chip. Each section of the SN74LS155 has a separate enable or control input.

A large decoder, such as the SN74LS154 4- to 16-line decoder provides you with a lot of address decoding flexibility. A single SN74LS154 decoder can be used to nonabsolutely decode 16 addresses and, when either IORQ* and either RD* or WR* is used to enable the decoder, the SN74LS154 will generate either 16-input device-select pulses or 16-output device-select pulses, with no additional gating required (Fig. 2-15).

A second decoder may be added to the circuit so that absolute device-select address pulses are generated directly. A typical example of this is shown in Fig. 2-16. Either RD* or WR* can be used to enable the lower decoder and IORQ* is used to enable the upper decoder. Note that only when 1010 is present on A7–A4 during an I/O operation will the 10 output of the upper decoder go to a logic 0. This signal, along with the RD* signal, is used to enable the lower decoder.

Thus, the device addresses are absolutely decoded with pulses for input devices A0H through AFH being generated. Of course, it would also be possible to take the same output from the upper decoder and use it, along with WR*, to enable another decoder. Thus, with three SN74LS154 decoders, it is possible to generate 16 absolutely decoded input device addresses and 16 absolutely decoded output device addresses. This should be more than enough for any interface,

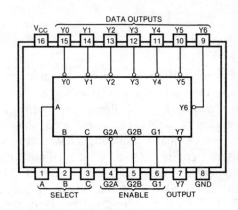

(A) Pin configuration.

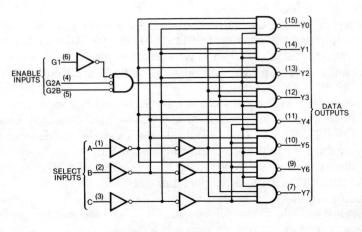

(B) Schematic diagram.

FIGURE 2-12

The SN74LS138 decoder integrated circuit.

since the interface is limited to a 4.5-inch × 6.5-inch (11.43 cm × 16.51 cm) card.

As you can see, it is very easy to use decoders to generate device-select pulses. There are literally hundreds of decoders to choose from depending on your application. The point that you need to remember is that they are easy to use, they provide a number of device-select pulses, and they have more flexibility than address-selection logic that is based on just gates.

USING COMPARATORS

Comparator-based address-decoding designs are relatively straightforward and they are very similar to the programmable-gate scheme that was shown in Fig. 2-4. Remember, comparators consist of a collection of gates, connected or integrated, to perform a comparison function. The comparator circuits allow us to preset an address that is constantly compared to an 8-bit accumulator I/O address or to a 16-bit memory address (for memory-mapped I/O). A typical comparator is the SN74LS85 4-bit magnitude comparator that is shown in Fig. 2-17. Besides the "equal" condition, the SN74LS85 can also detect the "greater than" and the "less than" condition, but these are not used in address-detection circuitry.

A typical address-comparison design is shown in Fig. 2-18, where the comparator has been set to 01101001, or 69H. Like an 8-input gate circuit, this scheme can only detect a single address, so most comparators are used with decoders for a flexible decoder design (Fig. 2-19).

The unique "equal" condition of the SN74LS85 is a logic 1, so it has to be inverted to be compatible with the logic 0 input requirement of the decoder's enable input. In this example, the decoder is only enabled when there is a match between addresses A7–A4 and the jumpers that are set to 0011, *and* the microprocessor is outputting information (IOWR*) to a peripheral device.

Again, the decoder is used to generate the absolutely decoded device-select pulses directly; in this case, the addresses for output devices 30H through 3FH. Of course, the inverted output of the comparator could also be used to enable a second SN74LS154 decoder, along with IORD*, so that address 30H through 3FH would be decoded for input devices.

MEMORY-MAPPED I/O

As you know already, the primary differences between accumulator I/O and memory-mapped I/O are that different control signals are used, and that a 16-bit memory address is placed on the address bus instead of an 8-bit address. As was the case with accumulator I/O, you have the choice of using either nonabsolutely decoded memory addresses or absolutely decoded memory addresses. Rather than have separate sections on decoder- and comparator-based address-detection circuitry, they will be combined.

The circuitry shown in Fig. 2-20 uses gates, comparators, and decoders. In this design, addresses A8–A15 have been wired to the eight inputs of a NAND gate. The inverted output of the NAND gate is used to enable an SN74LS85 comparator which compares the binary "values" of a set of four switches to the address on A4–A7. Finally, the output of the comparator (along with MEMRQ*) is used to enable an SN74LS154 4- to 16-line decoder. From this design, you can see that addresses A8–A16 must be logic 1s for the comparator to be enabled. This is fixed and cannot be changed by the user. Thus, memory addresses such as 11111111 01000101 and 11111111 01010111 will be decoded by this circuitry. However, the user can change the setting of the four DIP switches, so the user can choose the continuous block of 16 addresses that are decoded by the SN74LS154. The switch settings and the blocks that can be decoded are summarized in Table 2-5. Finally, the user can also select which of the decoder's outputs are used in the interface circuitry. Since only MEMRQ* is used to enable the decoder, the decoder's outputs have to be gated with either RD* or WR* before they are used to control I/O devices.

In this design, we have assumed that memory addresses between 11111111 00000000 and 11111111 11111111 (FF00H through FFFFH) can be dedicated to I/O devices. In 8085- and Z-80-based systems, this is reasonable, since the microprocessor is reset to location zero and interrupt vectors are near zero. However, in 6502- and 6800-based systems, the microprocessor is reset to HIGH memory, so there is the possibility that I/O device addresses will conflict with ROM memory locations.

The ideal design (if you can tolerate the high chip count) is a combination of three comparators and one decoder, where the upper twelve bits of the address (A4–A15) are detected with a combination of comparators and DIP switches, and the lowest four bits of the ad-

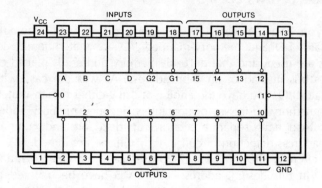

(A) Pin configuration.

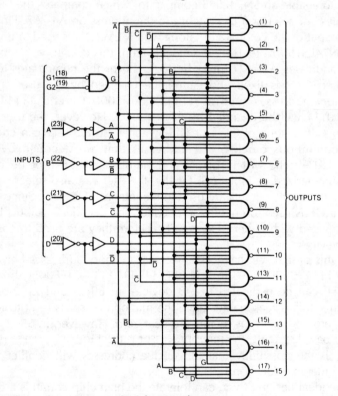

(B) Schematic diagram.

FIGURE 2-13
The SN74LS154 4- to 16-line decoder.

dress (A0–A3) are decoded by a decoder, such as the SN74LS154 (Fig. 2-21). From the design of the circuitry in Fig. 2-21, you can see that a combination of the comparator "chain" output and MEMRQ* is used to enable the decoder. Thus, the decoder outputs must be gated with either RD* or WR*. Then, these gated signals can be used to control peripherals.

The beauty of this design is the fact that it can be used with any CPU card in an STD bus system. As the amounts of memory and CPU cards are changed, the DIP switch positions can be changed.

Regardless of the address-detection circuitry used, you have to be careful that the memory addresses used by memory devices (EPROM, R/W memory, etc.) are not the same as the memory addresses used by peripherals. Thus, if your microcomputer contains 65,536 bytes of memory, it will be impossible for you to have 25 or 16 (or even one) memory-mapped I/O devices. On some CPU cards, some "tricks" are played that will violate this rule (MEMEX), but when you design your own peripheral cards, you should not violate this rule. At the same time, you cannot have two or more memory-mapped I/O devices that are activated when the same address is placed on the address bus. Thus, you cannot have three input devices that all use the address

INPUTS						OUTPUTS													
G1	G2	D	C	B	A	0	1	2	3	4	5	6	7	8	9	10	11	12	13	14	15
L	L	L	L	L	L	L	H	H	H	H	H	H	H	H	H	H	H	H	H	H	H
L	L	L	L	L	H	H	L	H	H	H	H	H	H	H	H	H	H	H	H	H	H
L	L	L	L	H	L	H	H	L	H	H	H	H	H	H	H	H	H	H	H	H	H
L	L	L	L	H	H	H	H	H	L	H	H	H	H	H	H	H	H	H	H	H	H
L	L	L	H	L	L	H	H	H	H	L	H	H	H	H	H	H	H	H	H	H	H
L	L	L	H	L	H	H	H	H	H	H	L	H	H	H	H	H	H	H	H	H	H
L	L	L	H	H	L	H	H	H	H	H	H	L	H	H	H	H	H	H	H	H	H
L	L	L	H	H	H	H	H	H	H	H	H	H	L	H	H	H	H	H	H	H	H
L	L	H	L	L	L	H	H	H	H	H	H	H	H	L	H	H	H	H	H	H	H
L	L	H	L	L	H	H	H	H	H	H	H	H	H	H	L	H	H	H	H	H	H
L	L	H	L	H	L	H	H	H	H	H	H	H	H	H	H	L	H	H	H	H	H
L	L	H	L	H	H	H	H	H	H	H	H	H	H	H	H	H	L	H	H	H	H
L	L	H	H	L	L	H	H	H	H	H	H	H	H	H	H	H	H	L	H	H	H
L	L	H	H	L	H	H	H	H	H	H	H	H	H	H	H	H	H	H	L	H	H
L	L	H	H	H	L	H	H	H	H	H	H	H	H	H	H	H	H	H	H	L	H
L	L	H	H	H	H	H	H	H	H	H	H	H	H	H	H	H	H	H	H	H	L
L	H	X	X	X	X	H	H	H	H	H	H	H	H	H	H	H	H	H	H	H	H
H	L	X	X	X	X	H	H	H	H	H	H	H	H	H	H	H	H	H	H	H	H
H	H	X	X	X	X	H	H	H	H	H	H	H	H	H	H	H	H	H	H	H	H

H = high level, L = low level, X = irrelevant

(C) Function table.

FIGURE 2-13 Cont.

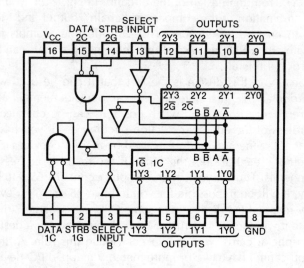

(A) Pin configuration.

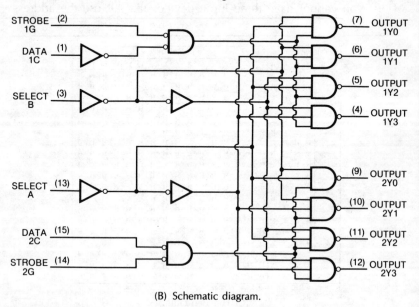

(B) Schematic diagram.

FIGURE 2-14
The SN74LS155 dual 2- to 4-line decoder.

B0FCH. One of the advantages of a Z-80- or 8085-based STD bus system is the fact that you *can* have 65,536 bytes of memory, and still have up to 256 input devices and 256 output devices. On 6502-, 6800-, and 6809-based systems, you can have 65,280 bytes of memory and, also, 256 input and 256 output devices.

Suppose that you are designing a peripheral card that will eventually become commercially available. Do you use accumulator I/O or memory-mapped I/O? Unfortunately, there is no strong consensus of opinion, one way or the other. Thus, if possible, the card should be designed to use either method, with the user selecting the "mode of operation." This may seem complex but, remember, the only difference between the two techniques is the control signals (MEMRQ* vs. IORQ*) and the size of the address (8 vs. 16 bits). In fact, the last comparator-decoder design that we illustrated can be *easily* modified to give us this capability.

The necessary modifications to the circuit are shown in Fig. 2-22. In the upper portion of the sketch, you can see the addition of an spst DIP switch (S1). S1 is used to switch the upper 8 bits of an address in

INPUTS			OUTPUTS			
SELECT	STROBE	DATA				
B A	1G	1C	1Y0	1Y1	1Y2	1Y3
X X	H	X	H	H	H	H
L L	L	H	L	H	H	H
L H	L	H	H	L	H	H
H L	L	H	H	H	L	H
H H	L	H	H	H	H	L
X X	X	L	H	H	H	H

INPUTS			OUTPUTS			
SELECT	STROBE	DATA				
B A	2G	2C	2Y0	2Y1	2Y2	2Y3
X X	H	X	H	H	H	H
L L	L	L	L	H	H	H
L H	L	L	H	L	H	H
H L	L	L	H	H	L	H
H H	L	L	H	H	H	L
X X	X	H	H	H	H	H

(C) Function tables for 2-line to 4-line decoder or 1-line to 4-line multiplexer.

FIGURE 2-14. Cont.

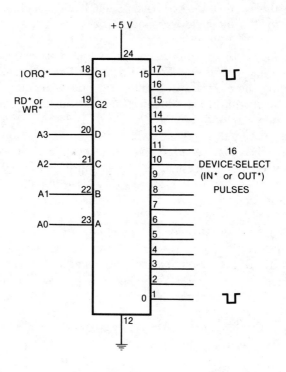

FIGURE 2-15

Using an SN74LS154 decoder to generate 16 nonabsolutely decoded (input or output) device-select pulses.

or out. Thus, if the switch is open, it is in the accumulator I/O setting (8 bits) and if it is closed, it's in the memory-mapped I/O setting (16 bits). A second switch (S2) is also required, along with some logic that is wired-in in the form of a 2-input multiplexer. This logic is used to select either IORQ* or MEMRQ* as the enable signal for the SN74LS154 decoder.

USING PROMS

Address-detection circuitry can also be designed and built using programmable read-only memories (PROMs). In essence, these devices are used (programmed) just like the gates and inverters that were discussed at the beginning of this chapter. Since PROMs are

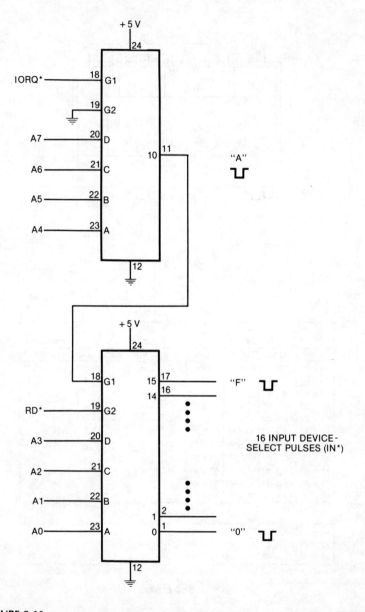

FIGURE 2-16
Absolutely decoding 16 device addresses using two SN74LS154 decoders.

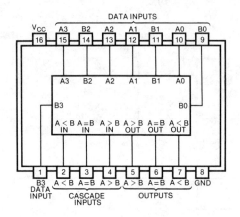

(A) Pin configuration.

COMPARING INPUTS				CASCADING INPUTS			OUTPUTS		
A3,B3	A2,B2	A1,B1	A0,B0	A>B	A<B	A=B	A>B	A<B	A=B
A3 > B3	X	X	X	X	X	X	H	L	L
A3 < B3	X	X	X	X	X	X	L	H	L
A3 = B3	A2 > B2	X	X	X	X	X	H	L	L
A3 = B3	A2 < B2	X	X	X	X	X	L	H	L
A3 = B2	A2 = B2	A1 > B1	X	X	X	X	H	L	L
A3 = B3	A2 = B2	A1 < B1	X	X	X	X	L	H	L
A3 = B3	A2 = B2	A1 = B1	A0 > B0	X	X	X	H	L	L
A3 = B3	A2 = B2	A1 = B1	A0 < B0	X	X	X	L	H	L
A3 = B3	A2 = B2	A1 = B1	A0 = B0	H	L	L	H	L	L
A3 = B3	A2 = B2	A1 = B1	A0 = B0	L	H	L	L	H	L
A3 = B3	A2 = B2	A1 = B1	A0 = B0	L	L	H	L	L	H

'85, 'LS85, 'S85

A3 = B3	A2 = B2	A1 = B1	A0 = B0	X	X	H	L	L	H
A3 = B3	A2 = B2	A1 = B1	A0 = B0	H	H	L	L	L	L
A3 = B3	A2 = B2	A1 = B1	A0 = B0	L	L	L	H	H	L

(B) Function tables.

FIGURE 2-17
The SN74LS85 4-bit magnitude comparator.

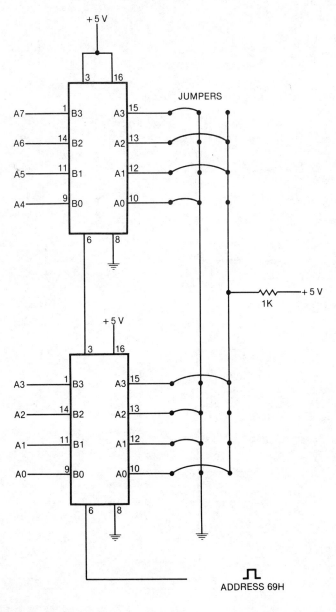

FIGURE 2-18
Using two SN74LS85 comparators to detect address 69H.

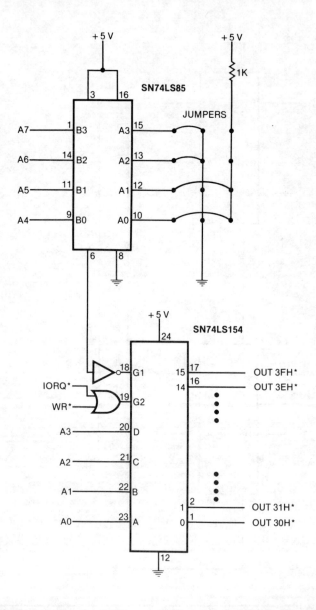

FIGURE 2-19
A flexible comparator-decoder approach for detecting I/O device addresses.

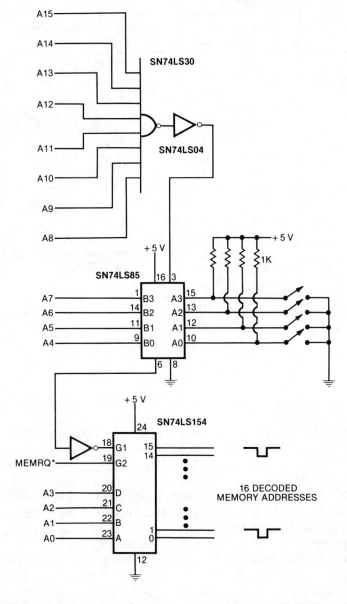

FIGURE 2-20
Absolutely decoding 16 memory addresses for I/O device synchronization.

Table 2-5. Decoded Blocks of Memory Addresses and Their Switch Settings

Switches				Memory Address Block
A7	A6	A5	A4	
0	0	0	0	FF00H–FF0FH
0	0	0	1	FF10H–FF1FH
0	0	1	0	FF20H–FF2FH
0	0	1	1	FF30H–FF3FH
0	1	0	0	FF40H–FF4FH
0	1	0	1	FF50H–FF5FH
0	1	1	0	FF60H–FF6FH
0	1	1	1	FF70H–FF7FH
1	0	0	0	FF80H–FF8FH
1	0	0	1	FF90H–FF9FH
1	0	1	0	FFA0H–FFAFH
1	0	1	1	FFB0H–FFBFH
1	1	0	0	FFC0H–FFCFH
1	1	0	1	FFD0H–FFDFH
1	1	1	0	FFE0H–FFEFH
1	1	1	1	FFF0H–FFFFH

1 = open; 0 = closed

used on many commercially available STD bus boards, it is important that they be discussed.

Since a PROM is a memory device, it contains a specific number of words and each word contains a specific number of bits. Thus, there are 256 × 4, 16 × 4 and 1024 × 4 PROMs (to mention just a few), where the first number is the number of words and the second number is the number of bits in each word. Since PROMs are digital devices, they all work with just 1s and 0s. Thus, there has to be a way that the user can specify one word among many. This is done by using the PROM's address lines.

Since address lines can assume either one of two binary states, there are 8 address lines (pins) for a 256 × 4 PROM, 4 for a 16 × 4 and 10 for a 1024 × 8. There is already a decoder built into the PROM that decodes this address and then selects one memory location. Knowing this, it should be easy for you to understand the func-

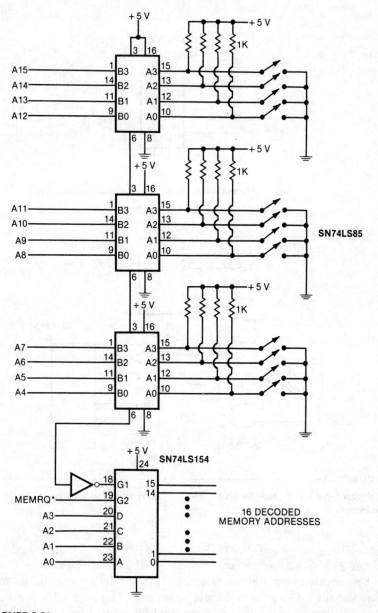

FIGURE 2-21
Using comparators and DIP switches to select any block of 16 memory addresses.

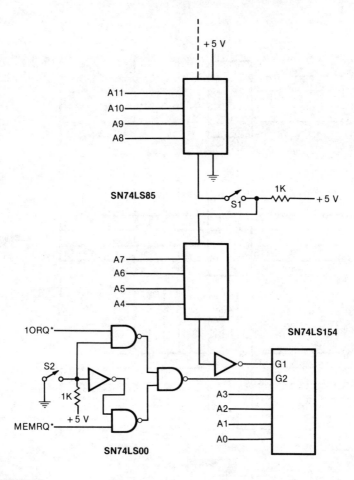

FIGURE 2-22
A decoder that can be used to decode either memory addresses or accumulator I/O addresses.

tion of the pins for the HM-76LS03 32 × 8 PROM that is illustrated in Fig. 2-23.

Before discussing address-detection circuitry, let's use a PROM to simulate the function of a 4-input NAND gate. The truth table for the device is shown in Table 2-6. From the truth table, you can see that the only time that the output is a logic 0 is when all of the inputs are

Table 2-6. Truth Table for an SN74LS20 4-Input NAND Gate

Inputs				Output
A	B	C	D	Q
0	0	0	0	1
0	0	0	1	1
0	0	1	0	1
0	0	1	1	1
0	1	0	0	1
0	1	0	1	1
0	1	1	0	1
0	1	1	1	1
1	0	0	0	1
1	0	0	1	1
1	0	1	0	1
1	0	1	1	1
1	1	0	0	1
1	1	0	1	1
1	1	1	0	1
1	1	1	1	0

logic 1s. Now, let's examine the *unprogrammed* contents of an imaginary 16 × 1 PROM (Table 2-7). As you can see from Table 2-7, regardless of the memory address used, the PROM contains a logic 1 in each memory location. However, PROMs are programmable devices. This means that either 1s or 0s can be programmed into the device, depending on the device used. Since our 16 × 1 PROM contains all 1s in the unprogrammed state, we must be able to program 0s into it.

What must be done to make the PROM look like an SN74LS20 4-input NAND gate? The only operation that must be performed is the programming of a logic 0 into memory location 15_{10} (1111, in binary). This means that the address inputs of the PROM are treated just like the inputs to the NAND gate, and the output of the PROM is treated like the output of the NAND gate.

As you might guess, the 16 × 1 PROM can be programmed to simulate the operation of *any* 4-input gate, or any combination of gates that has four inputs and one output! Using a new 16 × 1 PROM, a 4-input OR gate would be obtained by programming a 0 only in location 0, and a 4-input AND gate would be obtained by programming 0s in all memory locations, except for memory location

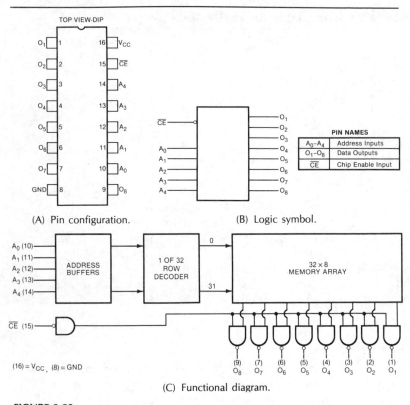

FIGURE 2-23
The HM-76LS03 32 × 8 PROM. *(Courtesy Harris Semiconductor Products)*

15_{10} (1111, in binary). Unfortunately, if you make a mistake when programming PROMs, you cannot repair the mistake. Instead, you have to throw the PROM out and start over again.

Using a 256 × 4 PROM, four accumulator I/O addresses can be absolutely decoded. However, by wiring it in a slightly different manner, four device-select pulses can be generated. The pin diagram for the SN74S287 256 × 4 PROM is shown in Fig. 2-24. This is the device that we will use in the examples. Address input AD A is the least-significant address bit and input AD H is the most-significant address bit. The two enable inputs, $\overline{S1}$ and $\overline{S2}$, must both be logic 0s for the PROM to be enabled.

Suppose that you need to absolutely decode addresses F0H, CBH, 09H, and 54H on the same card. Could you do this with the 256 × 4

PROM? Absolutely. This would be very difficult, however, if you were using the gate decoder-comparator circuitry discussed previously. A circuit that can be used is shown in Fig. 2-25. As you can see, the PROM is only enabled when a programmed I/O operation is taking place. The address on the address bus is then used to select one of the 256 memory locations contained in the PROM. Only when address F0H, CBH, 54H, or 09H is on the address bus, should the appropriately labeled output go to the logic zero state. Thus, in memory location F0H in the PROM, a 1110 would be programmed, assuming that D01 is the least-significant bit (LSB) and that D04 is the most-significant bit (MSB). In location CBH, a 1101 would be programmed, in 54H, a 1011, and in 09H, a 0111 would be programmed. This assumes that the PROM contained all 1s in the unprogrammed state.

It is also possible to program the PROM so that the active state is a logic 1. In order to do this, all the memory locations, *except* the one that you are interested in, would be programmed with 0s. The one memory location that you are interested in would be left unprogrammed, or in the logic one state.

If a design such as the one shown in Fig. 2-25 is used, the decoded address pulses still have to be gated with either IORD* or IOWR*.

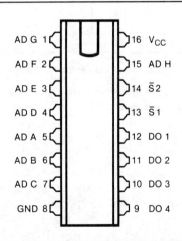

FIGURE 2-24

Pin configuration for an SN74S207 256 × 4 PROM. *(Courtesy Texas Instruments Incorporated)*

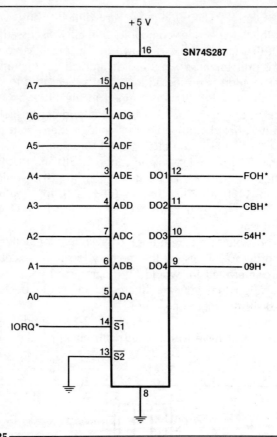

FIGURE 2-25

Absolutely decoding four accumulator I/O device addresses using a 256 × 4 PROM.

Also, some PROMs contained all 0s in the unprogrammed state. If you need an output that is active in the logic 1 state, just program a 1 in the location of interest. If you need an output that is active in the logic 0 state, you will have to program 1s in all of the locations other than the one of interest.

One final note about PROMs is that, if, as you are looking over an STD bus card, you find an IC that doesn't have a "standard" number, or which has been marked with a special manufacturer or "house" number, there is a very good chance that it is a PROM. Unless you take the PROM out and cycle it through all the addresses and generate a truth table, you will have to take it on faith that the PROM is working properly.

Table 2-7. Truth Table for an Imaginary 16 × 1 PROM

| Memory Address | | | | Contents of |
D	B	C	A	Memory
0	0	0	0	1
0	0	0	1	1
0	0	1	0	1
0	0	1	1	1
0	1	0	0	1
0	1	0	1	1
0	1	1	0	1
0	1	1	1	1
1	0	0	0	1
1	0	0	1	1
1	0	1	0	1
1	0	1	1	1
1	1	0	0	1
1	1	0	1	1
1	1	1	0	1
1	1	1	1	1

CONCLUSION

In this chapter, you have seen how device-address decoders are designed. This is important because they have to be designed into every peripheral design, and you may see some of these designs on boards that you purchase. Also, you can use these designs on your own boards. Depending on the type of I/O being used, either an 8- or 16-bit address has to be decoded. If accumulator I/O is used, then an 8-bit address is decoded. If memory-mapped I/O is used, a 16-bit address must be decoded. On many boards, the manufacturer will have provided the circuitry so that either of these I/O techniques may be used.

The actual decoder circuitry may be based on gates, decoders, comparators, or even a type of read-only memory (PROM). In conjunction with these different chips, the user may set the I/O address either through a set of soldered jumpers or by changing the setting of some DIP switches.

Once the address is decoded, the device address must be gated with the proper combination of control signals. If accumulator I/O is

Output Port Interfacing
Chapter 3

In Chapter 2, we discussed the methods used to detect an address on the address bus during either an I/O operation or a memory operation. Using gates, decoders, comparators, PROMs, or a combination of these devices, the user-specified address (specified in the hardware by adding jumpers or setting DIP switches) was "compared to" the address on the address bus. When a "match" occurred, a pulse was generated. In many of the circuits, the IORQ* or MEMRQ* signal was used, along with RD* and/or WR* to "qualify" the signal. The result, which is a device-select pulse, can then be used to control a peripheral.

Of course, if a computer can *only* generate pulses, there won't be many computers around, because they won't be very useful. Thus, while control pulses are important, the computer also has to have the ability to transfer data. In this chapter, we will be discussing the interface techniques used when the microcomputer needs to *output* information to peripherals using *output ports*.

OUTPUT TIMING

As you know from reading the previous two chapters, it is very important that you understand the timing of the I/O processes; that is, the relationships between the various control signals that the CPU card produces. As you discovered in Chapter 1, a knowledge of the timing of a system can explain why one "family" chip won't work with a different microprocessor chip or CPU card.

When data is placed on the data bus by the microprocessor, regardless of the microprocessor used in your STD bus system, it will only be on the data bus for a very short period of time—typically from 300 to 700 nsec. Thus, it is the responsibility of the memory and interface circuitry to latch or "grab" this information off of the data bus at the appropriate time. The Write timing specified by Mostek for their STD-Z80 CPU cards is shown in Fig. 3-1; the times are summarized in Table 3-1. As you can see, the data is only on the data bus for about 700 nsec during a Write operation. Unfortunately, most peripherals cannot respond to data that is only on the data bus for such a short period of time.

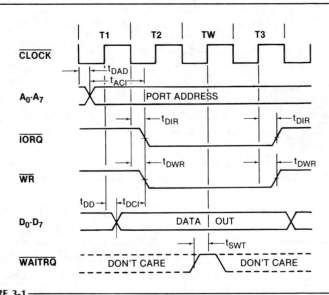

FIGURE 3-1

IOWR* timing in an STD-Z80 bus system. *(Courtesy Mostek Corp.)*

Let's assume that an 8-bit value being output by the microcomputer represents the position of a 10-ton hydraulic ram. A value of 0 means that the ram is retracted all the way in, and a value of 255_{10} means that the ram is extended 2 feet. As you can imagine, it will take a certain amount of time (possibly, from 2–3 seconds to 2–3 minutes) to move the ram as new positional values are output by the computer. Thus, the hydraulic-ram interface has to latch, or remember the value, that was on the data bus for 700 nsec.

Table 3-1. Timing Specifications for the STD-Z80 Bus (4 MHz)

BUS SIGNAL	SYMBOL	DESCRIPTION	MIN	MAX	UNITS
/CLOCK	t_c	Clock period.		250	nsec
A_0-A_{15}	t_{DAD}	Address output delay.		130	nsec
	t_{ACM}	Address stable prior to /MEMRQ.	75		nsec
	t_{ACI}	Address stable prior to /IORQ, /RD, /WR (I/O cycle).	180		nsec
D_0-D_7	t_{SDO}	Data setup time during /M1 cycle.	70		nsec
	t_{SDR}	Data setup time during memory-read cycle, I/O read cycle, or interrupt acknowledge cycle.	80		nsec
	t_{DCM}	Data stable prior to /WR (memory write)	80		nsec
	t_{DD}	Data output delay.	155		nsec
	t_{DCI}	Data stable prior to /WR (I/O write).	−30		nsec
	t_{CDF}	Data stable from /WR.	70		nsec
/MEMRQ	t_{DMR}	/MEMRQ delay from /CLOCK.		95	nsec
	t_{MRH}	/MEMRQ pulse width high.	105		nsec
	t_{MRL}	/MEMRQ pulse width low.	220		nsec
/IORQ	t_{DIR}	/IORQ delay from /CLOCK.		85	nsec
/RD	t_{DRD}	/RD delay from /CLOCK.		95	nsec

Table 3-1. Cont. Timing Specifications for the STD-Z80 Bus (4 MHz)

BUS SIGNAL	SYMBOL	DESCRIPTION	MIN	MAX	UNITS
/WR	t_{DWR}	/WR delay from /CLOCK.		75	nsec
/M1	t_{DMI}	/M1 delay from /CLOCK.		110	nsec
/REFRESH	t_{DRF}	/REFRESH delay from /CLOCK.		130	nsec
/WAITRQ	t_{SWT}	/WAITRQ setup time.	85		nsec
/INTRQ	t_{SINT}	/INTRQ setup time.	95		nsec
/NMIRQ	t_{SNMI}	/NMIRQ setup time.	85		nsec
/BUSRQ	t_{SBR}	/BUSRQ setup time.	55		nsec
/BUSAK	t_{DBA}	/BUSAK delay from /CLOCK.		110	nsec
	t_{TS}	Tri-state buffer delay from /BUSAK for A_0-A_{15}, D_0-D_7, /MEMRQ, IORQ, /M1, /RD, /WR, /REFRESH, and /INTAK.		25	nsec

Courtesy Mostek Corp.

The type of circuit that can perform this function is called a *latch*, since it can take information off of the data bus and hold it until new information is output to the circuit, or until power is lost. Like the decoder, gate, and comparator chips discussed in Chapter 2, there is an endless variety of latch chips, each with different amounts of input, output, and control pins. In fact, there are also a number of counter, decoder, shift register, and communications chips that also contain latches. Typical latch chips are summarized in Table 3-2 and some of the other chips that contain latches are summarized in Table 3-3.

LATCHES IN OUTPUT PORTS

The pin configurations and function tables for the SN74LS75, SN74LS373, and SN74LS175 integrated circuits are shown in Fig. 3-2. In examining the function table for the SN74LS373, you can see that whenever the enable (G) input is a logic one, the data or logic level present on one of the "D" inputs is passed through the latch to the appropriate "Q" output. When the enable input goes from a logic one to a logic zero, the level present on the D input at this time is "latched," or remembered. The timing relationships between a data input, the enable input, and an output pin are shown in Fig. 3-3. As you can see, as soon as the "G" input goes to a logic one, the Q output assumes the state of the D input, even when the levels on the D input change. The logic levels on the D input are present on the Q output as long as the "G" input remains in the logic one state. When the "G" input goes to a logic zero, whatever value is present on the D input will be latched or remembered by logic contained in the chip, and this same logic level will be output on the Q output. The Q output will remain in this state until either the "G" input goes to a logic one or power is lost.

Notice, also, that the SN74LS373 has another control pin that is labeled *output control*. When an SN74LS373 is used in an output port design, this pin is simply wired to ground. In the next chapter, you will see how this pin can be used.

How would a device such as the SN74LS373 be used as an output port? The D inputs to the device would be wired to the data bus and

Table 3-2. Typical Latching Integrated Circuits

Device Number	Pin Count	Bits Latched
SN74LS75	16	4
SN74LS100	24	8
SN74LS174	16	6
SN74LS373	20	8

Device Number	Control Inputs	Bits Output
SN74LS75	2	2 sets of 4
SN74LS100	2	8
SN74LS174	2	6
SN74LS373	2	8

Table 3-3. Integrated Circuits That Contain Latches

Device Number	Function
SN74LS94, 95	4-bit shift registers
SN74LS165, 166	8-bit shift registers
SN74LS190–193	4-bit binary and BCD counters
SN74LS199	8-bit shift register
SN74LS278	4-bit priority register
SN74LS298	Quad 2-input multiplexer
SN74142–144	Counter/latch/decoder-driver

the "G" input would be wired to a device-select pulse (a combination of an address-detection pulse, IORQ* and WR*). The latched outputs of the device could then be wired to a peripheral device. The device might be as simple as eight individual LEDs, or two 7-segment decoder/drivers for use with two 7-segment LED displays, or a floppy disk.

A typical output port that is based on the SN74LS373 is shown in Fig. 3-4. In this design, the light-emitting diodes are used to indicate the logic 1s and 0s that are latched. The SN74LS05 inverters are used to buffer the outputs of the latches and sink the current that is required to drive the LEDs. Some simple software that can be used with this interface is shown in Example 3-1 for some of the STD bus processors.

Since you are probably already familiar with the assembly language instruction set of your STD bus processor, we will not go into a lot of details about this software. However, the programs simply increment a register and output this information to the output port. Since the microcomputer is very fast, the register can be incremented very fast and we would not see the state of the LEDs change. Therefore, the microprocessor executes a time-delay loop 256 times, which increases the time it takes for the microprocessor to increment the register and output it and, then, do it again.

In this software, we have also assumed that the output port is output port 00H. In the 8085 and Z-80 microprocessors, this address is actually contained in the OUT instruction. Since the 6800 and 6502 processors only "think" that they can do memory-mapped I/O, we have assumed that whenever these two processors are communicating with "memory locations" C000H through C0FFH, they are really communicating with "accumulator" I/O devices. At this time, we

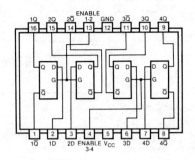

(A) SN74LS75.

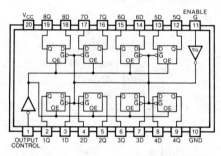

(B) SN74LS373.

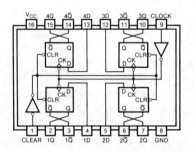

(C) SN74LS175.

FIGURE 3-2
Some pin configurations and function tables.

should note that we have not tried to "tweak" every last byte or μsec out of the program. We are not writing benchmarks, we are just using some simple examples to give you some idea of what the I/O software looks like.

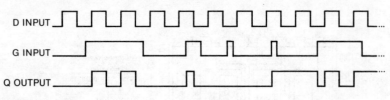

FIGURE 3-3
Latch timing relationships for the SN74LS373.

Table 3-4. Light Assignments of Bits for Main and Elm Streets

Data Bit	Function
D0	Red on Main
D1	Yellow on Main
D2	Green on Main
D3	Red on Elm
D4	Yellow on Elm
D5	Green on Elm
D6	Not used
D7	Not used

A TRAFFIC LIGHT CONTROLLER

At this point, you may be wondering what a "real" interface looks like because nothing can be this simple. However, a lot of peripherals are this simple once you break them down into address-detection logic, output ports, and input ports (see Chapter 4). In fact, the output port shown in Fig. 3-4 can be modified to control a traffic light. Instead of using LEDs, you would use the output port to drive a solid-state relay (SSR), which would turn the 110-V 60-W ac light bulbs in the traffic light on and off.

The new circuit is shown in Fig. 3-5. Of course, before any software could be written, we would have to know which bit in the output port turns on which light in the traffic light. This information is contained in Table 3-4.

At this point, you have to decide exactly what you want the STD bus computer to do. To keep the interface and the software simple, the computer will be used as a simple timer. We won't worry about the sensors in the street that sense vehicles and, then, after a certain count, switch the lights. Instead, we need to let traffic pass through

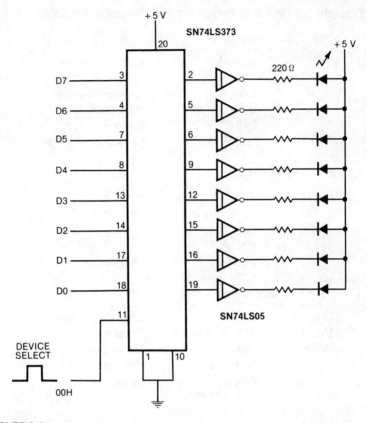

FIGURE 3-4
An LED output port based on the SN74LS373 latch.

the light on Main Street for 30 seconds, followed by traffic flowing on Elm Street for 10 seconds. In order to do this, we have to determine the bit patterns that will turn the appropriate combinations of lights on and off. There are really only four combinations of 1s and 0s that we will need: (1) green on Main, red on Elm, (2) yellow on Main, red on Elm, (3) red on Main, green on Elm, and (4) red on Main, yellow on Elm. The bit patterns that will cause these combinations of lights to be activated are summarized in Table 3-5.

In our software planning, we will need to store this *table* of information in memory, and then periodically access and output a value to the output port that controls the solid-state relays (SSRs). A time-

Example 3-1. Ways to Increment and Output a Register.

1. For an 8085 processor:

START:	XRA A	;Clear A register
START1:	OUT 00H	;Output A register
LOOP:	INR B	;Increment count
	JNZ LOOP	;Wait until zero
	INR A	;Increment A
	JMP START1	;Output new value

2. For a Z-80 processor:

START	XOR A	;Clear A
START1	OUT (00H),A	;Output it
LOOP	DJNZ LOOP	;Decrement B
	INC A	;Increment A
	JR START1	;Output A

3. For a 6502 processor:

START	LDY #$00	;Clear Y to zero
START1	STY $C000	;Output it
LOOP	INX	;Increment X
	BNE LOOP	;Wait until zero
	INY	;Increment Y
	JMP START1	;Output new value

4. For a 6800 processor:

START	CLR A	;Clear A
START1	STA $C000	;Output it
LOOP	INC B	;Increment B
	BNE LOOP	;Wait
	INC A	;Increment A
	BRA START1	;Output it

delay subroutine which will generate a 1-second delay will also be needed in the software. This subroutine will have to be called a number of times in order to generate the delays required between the changing of the bit pattern on the output port. The constants in the time-delay subroutine may have to be adjusted, depending on the

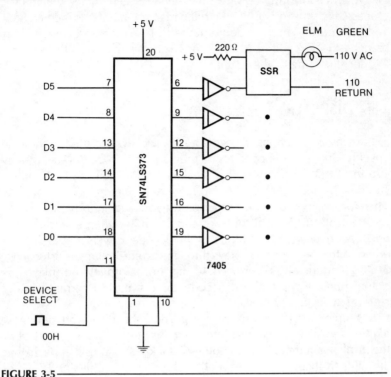

FIGURE 3-5
Using an IC output port to control a traffic light.

cycle time of the microprocessor that is used on your STD bus CPU card.

In Examples 3-2 through 3-5, the microprocessors first get the address of the list that contains the patterns and the time-delay values. In LOOP, the pattern is fetched from memory and output to the traffic light and, then, the time delay is fetched from the same list. The DELAY subroutine is then called which, in turn, calls a 1-second time

Table 3-5. Bit Combinations for Controlling the Traffic Light

Data Byte	Function
00001100	Green on Main, red on Elm
00001010	Yellow on Main, red on Elm
00100001	Red on Main, green on Elm
00010001	Red on Main, yellow on Elm

delay. Once the appropriate time delay is generated, the microprocessor returns to LOOP where the memory pointer is incremented and the count of 4 is decremented. The microprocessor then branches back or jumps to LOOP or TRAF, depending on the result of the decrementing of the count.

LED DISPLAYS

In many applications, 7-segment LED displays are used to indicate specific values to the operator or programmer. These devices usually use the BCD or Binary-Coded Decimal code (Table 3-6), so only 4 bits of information are required to represent the numbers 0 through 9. Therefore, using our 8-bit output port, two 7-segment displays could be controlled. The outputs of the latch would be wired to 2 *decoder/drivers* and the outputs of the decoder/drivers would then be wired to the 7-segment displays through current-limiting resistors (Fig. 3-6). Unfortunately, if we need 10 digits of information displayed, 5 output ports, along with 10 decoder/drivers and 70 current-limiting resistors, will be required.

A simpler interface can be designed if the displayed output is *multiplexed*. This means that information is only output to one digit in the display at a time, but it is done so quickly that to us it appears that all digits in the display are on all of the time. Multiplexing LED displays is very common and is done in calculators, digital watches, and digital clocks.

Using one 8-bit output port, up to 16 digits in a display can be controlled. As you can see in Fig. 3-7, the interface for our 10-digit multiplexed LED display is very simple. In this interface, the DS8857 decoder/driver decodes the 4-bit BCD code onto bits D4–D7 of the output port and *sends* current to the appropriate combination of segments in *all of the digits in the display*. The SN7442 decoder decodes the information on bits D0–D3 of the output port and selects which digit will be turned on, since the SN7442 acts like a current sink to ground. By changing the 7-segment value and the digit-enable value quickly enough, all of the displays will appear to be on, all of the time.

How often does the information latched in the output port have to be changed, in order to make all of the digits appear to be on? This depends on the number of digits used. For a 10-digit display, each digit would have to be turned on and off about 50 times per second.

Example 3-2. Simple Traffic-Light Software for the 8085

TRAF:	LXI H,PATTRN	;Get table address
	MVI C,04H	;Number of entries
LOOP:	MOV A,M	;Get entry
	OUT 00H	;Output it
	INX H	;Increment address
	MOV A,M	;Get time delay value
	CALL DELAY	;Delay
	INX H	;Get to next entry
	DCR C	;Done all 4?
	JNZ LOOP	;No, do another
	JMP TRAF	;Yes, start over
DELAY:	CALL D1SEC	;Delay 1 sec
	DCR A	;Decrement loop count
	RZ	;Done? Yes, return
	JMP DELAY	;No, wait 1 sec
D1SEC:	PUSH PSW	;Save A and flags
	MVI B,0AH	;B = decimal 10
D1:	LXI D,1E00H	;100 ms value
D11:	DCX D	;Decrement value
	MOV A,D	;Get part
	ORA E	;OR with part
	JNZ D11	;Wait some more
	DCR B	;100 ms gone, loop?
	JNZ D1	;Yes, do another
	POP PSW	;Get A and flags
	RET	
PATTRN;	0CH	;Green MAIN, Red ELM
	1EH	;30 sec
	0AH	;Yellow MAIN, Red ELM
	05H	;5 sec
	21H	;Red MAIN, Green ELM
	0AH	;10 sec
	11H	;Red MAIN, Yellow ELM
	05H	;5 sec

Example 3-3. Simple Traffic-Light Software for the Z-80

TRAF	LD HL,PATTRN	;Get table address
	LD C,04H	;Number of entries
LOOP	LD A,(HL)	;Get entry
	OUT (00),A	;Output it
	INC HL	;Increment address
	LD A,(HL)	;Get time delay value
	CALL DELAY	;Delay
	INC HL	;Get to next entry
	DEC C	;Done all 4?
	JR NZ,LOOP	;No, do another
	JR TRAF	;Yes, start over
DELAY	CALL D1SEC	;Delay 1 sec
	DEC A	;Decrement loop count.
	RET Z	;Done? Yes, return
	JP DELAY	;No, wait 1 sec
D1SEC	PUSH AF	;Save A and flags
	LD B,0AH	;B = decimal 10
D1	LD DE,1E00H	;100 ms value
D11	DEC DE	;Decrement value
	LD A,D	;Get part
	OR E	;OR with part
	JR NZ, D11	;Wait some more
	DEC B	;100 ms gone, loop?
	JR NZ, D1	;Yes, do another
	POP AF	;Get A and flags
	RET	
PATTRN:	0CH	;Green MAIN, Red ELM
	1EH	;30 sec
	0AH	;Yellow MAIN, Red ELM
	05H	;5 sec
	21H	;Red MAIN, Green ELM
	0AH	;10 sec
	11H	;Red MAIN, Yellow ELM
	05H	;5 sec

Example 3-4. Simple Traffic-Light Software for the 6502

TRAF	LDX #$00	;Index to zero
	LDY #$04	;4 table entries
LOOP	LDA PATTRN,X	;Get an entry
	STA $C000	;Output it
	INX	;Increment the index
	LDA PATTRN,X	;Get time delay
	JSR DELAY	;Delay for a while
	INX	;Get to another pattern
	DEY	;Decrement count
	BNE LOOP	;Done all 4 combos?
	JMP TRAF	;Yes, do them again
DELAY	STA $20	;Save 1 sec count
DLY1	JSR D1SEC	;Delay 1 sec
	DEC $20	;Decrement count
	BNE DLY1	;Not zero, do again
	RTS	;All done
D1SEC	LDA #$0A	;Decimal 10
	STA $21	;Save it
D100MS	LDA #$00	;Save a 16-bit
	STA $22	;time constant
	LDA #$31	;in memory.
	STA $23	
DLOOP	DEC $22	;Decrement LSBs
	BNE DLOOP	
	DEC $23	;Decrement MSBs
	BNE DLOOP	
	DEC $21	;Done 100 ms, do
	BNE D100MS	;again?
	RTS	
PATTRN:	0CH	;Green MAIN, Red ELM
	1EH	;30 sec
	0AH	;Yellow MAIN, Red ELM
	05H	;5 sec
	21H	;Red MAIN, Green ELM
	0AH	;10 sec
	11H	;Red MAIN, Yellow ELM
	05H	;5 sec

Example 3-5. Simple Traffic-Light Software for the 6800

TRAF	LDX PATTRN	;Index to zero
	LDA B #$04	;4 table entries
LOOP	LDA A 0,X	;Get an entry
	STA A $C000	;Output it
	INX	;Increment the index
	LDA 0,X	;Get time delay
	JSR DELAY	;Delay for a while
	INX	;Get to another pattern
	DEC B	;Decrement count
	BNE LOOP	;Done all 4 combos?
	JMP TRAF	;Yes, do them again
DELAY	STA A $30	;Save 1 sec count
DLY1	JSR D1SEC	;Delay 1 sec
	DEC $0030	;Decrement count
	BNE DLY1	;Not zero, do again
	RTS	;All done
D1SEC	LDA A #$0A	;Decimal 10
	STA A $31	;Save it
D100MS	LDA A #$00	;Save a 16-bit
	STA A $32	;time constant
	LDA A #$31	;in memory.
	STA A $33	
DLOOP	DEC $0032	;Decrement LSBs
	BNE DLOOP	
	DEC $0033	;Decrement MSBs
	BNE DLOOP	
	DEC $0031	;Done 100 ms, do
	BNE D100MS	;again?
	RTS	
PATTRN:	0CH	;Green MAIN, Red ELM
	1EH	;30 sec
	0AH	;Yellow MAIN, Red ELM
	05H	;5 sec
	21H	;Red MAIN, Green ELM
	0AH	;10 sec
	11H	;Red MAIN, Yellow ELM
	05H	;5 sec

Table 3-6. The Binary-Coded Decimal (BCD) Code

BCD	Decimal
0000	0
0001	1
0010	2
0011	3
0100	4
0101	5
0110	6
0111	7
1000	8
1001	9
1010	Not valid
1011	Not valid
1100	Not valid
1101	Not valid
1110	Not valid
1111	Not valid

Thus, new information would have to be output to the output port about every 2 msec (1 second/(50 × 10)).

The software that perform this action is listed in Examples 3-6 through 3-9. The information that is displayed is stored in ten different memory locations, starting at DATA. However, the information is really stored in only the four least-significant bits (LSBs) of each memory location (in BCD). A packed-BCD format could also be used, where two digits are stored in each memory location, but only five storage locations would be saved, and five or more additional locations would probably have to be used to store instructions to unpack the digits.

In general, the microprocessor has to set up some pointers and a count of 10_{10}. The BCD information is read from memory, rotated or shifted to the left four times so that it is in the 4 most-significant bits (MSBs), and, then, a digit-enable value between 0 and 9 is added to this. The result is written out to the display. The microprocessor spends about 2 msec in a loop, so that the information is displayed for 2 milliseconds. The microprocessor then increments an index or pointer and decrements the count. If the count is nonzero, the next digit is fetched and processed. Otherwise, the microprocessor starts from the beginning again.

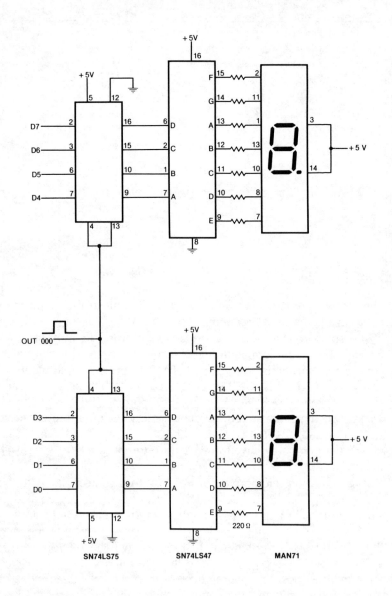

FIGURE 3-6
A latched 2-digit LED display.

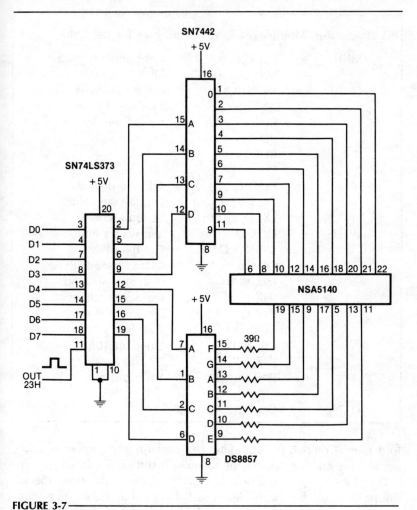

FIGURE 3-7

A 10-digit multiplexed LED display interface.

One problem with this interface, and the software, is the fact that the display requires constant attention from the microprocessor. This means that the microprocessor cannot process data for 32 seconds and, then, output the information once so that the operator can see it and continue processing. The display must be updated and it must be updated every 2 msec. The advantage of a "static" display that uses lots of latches, decoder/drivers, and resistors is the fact that once in-

99

Example 3-6. Multiplexed Display Software for the 8085

MP10:	LXI H,DATA	;Get pointer to data
	MVI C,00H	;Digit is zero
DIGIT:	MOV A,M	;Get a data value
	ANI 0FH	;Save the 4 LSBs
	RLC	;Move them to the
	RLC	;MSBs.
	RLC	
	RLC	
	ADD C	;Add the digit #
	OUT 23H	;Display the value
	LXI D,0100H	;Set up a count
WAIT:	DCX D	;and wait a little.
	MOV A,D	;Get the MSBs
	ORA E	;OR with the LSBs
	JNZ WAIT	;Not zero, so loop
	INX H	;Increment address
	INR C	;Increment digit #
	MOV A,C	;Get digit #
	CPI 0AH	;Done all 10?
	JNZ DIGIT	;No, do another
	JMP MP10	;Yes, do again
DATA:	DB 1,2,3,4,5,6,7,8,9	

formation is output, the computer can perform any number of additional tasks and the same information will still be displayed.

Because multiplexed displays require constant attention, special display control chips were developed. They are available from Intel, NEC, National Semiconductor, and Matrox. The pin configuration, truth tables, and block diagram for the MM74C917 6-Digit Hex Display Controller/Driver is given in Fig. 3-8.

To the microprocessor, this device appears to be 8 output ports, all with consecutive addresses. From the truth table, you can see that the "devices" at addresses 6 and 7 are really null or "dummy" digits (which means that they do not exist). Thus, each one of these devices will waste two device addresses.

To the displays, this device appears to be multiplexing logic, so a large number of decoder/drivers and resistors are not necessary. The

Example 3-7. Multiplexed Display Software for the Z-80

```
MP10      LD HL,DATA      ;Get pointer to data
          LD C,00H        ;Digit is zero
DIGIT     LD A,(HL)       ;Get a data value
          AND 0FH         ;Save the 4 LSBs
          RLCA            ;Move them to the
          RLCA            ;MSBs.
          RLCA
          RLCA
          ADD C           ;Add the digit #
          OUT (23H),A     ;Display the value
          LD DE,0100H     ;Set up a count
WAIT      DEC DE          ;and wait a little.
          LD A,D          ;Get the MSBs
          OR E            ;OR with the LSBs
          JR NZ,WAIT      ;Not zero, so loop
          INC HL          ;Increment address
          INC C           ;Increment digit #
          LD A,C          ;Get digit #
          CP 0AH          ;Done all 10?
          JR NZ,DIGIT     ;No, do another
          JR MP10         ;Yes, do again

DATA:     DB 1,2,3,4,5,6,7,8,9
```

only additional circuitry that is required along with this chip are 8 current-limiting resistors and 6 npn transistors that sink the current from each one of the digits in the display. Because of this, only common-cathode LED displays can be used with this device.

An interface based on this display controller chip is shown in Fig. 3-9. The only unusual feature of this interface is the fact that the address-detection logic *does not* generate pulses such as 40H, 41H, 42H, 43H, 44H, and 45H. That is, there are no individual device-select pulses generated. These pulses are not generated for the simple reason that the controller chip contains six (really eight) output ports. Thus, the controller chip decodes the state of address lines A0–A2 and the address-detection logic has to detect the proper address on A3–A7.

The software for writing six BCD values out to this device is given

Example 3-8. Multiplexed Software for the 6502

MP10	LDA #$00	;Digit is zero
	STA $30	;Save it
	TAX	;Use as an index
DIGIT	LDA DATA,X	;Get a data value
	ASL	;Move it to the
	ASL	;4 MSBs.
	ASL	
	ASL	
	CLC	;Clear the carry
	ADC $30	;Add the digit #
	STA $C023	;Output value
	LDA #$A0	;Load a delay value
WAIT	SBC #$01	;Subtract 1
	BNE WAIT	;Loop until zero
	INX	;Increment index
	INC $30	;Increment digit #
	LDA #$0A	
	CMP $30	;Done all 10?
	BNE DIGIT	
	JMP MP10	
DATA:	DB 1,2,3,4,5,6,7,8,9	

in Examples 3-10 and 3-11. All this software has to do is transfer the BCD numbers (stored in consecutive memory locations) out to the controller chip. The Z-80's software is the simplest, since one of its *block I/O instructions* can be used. The instruction used (OUTI) outputs the 8-bit content of a memory location, increments the memory address, and decrements a count. Because the I/O address has to be incremented (from 20H up to 25H), one of the I/O repeat instructions cannot be used.

DIGITAL-TO-ANALOG CONVERTERS

Digital-to-analog converters (DACs) are used in many applications. The STD bus user can purchase cards that contain one or more of these devices from a number of manufacturers, or the user can pur-

Example 3-9. Multiplexed Software for the 6800

MP10	LDX DATA	;Digit is zero
	CLR B	;Save it
DIGIT	LDA A 0,X	;Use as an index
	ASL	;Get a data value
	ASL	;Move it to the
	ASL	;4 MSBs.
	ASL	
	ABA	
	STA $C023	;Clear the carry
	LDA A #$80	;Add the digit #
WAIT	DEC A	;Output value
	BNE WAIT	;Load a delay value
	INX	;Subtract 1
	INC B	;Loop until zero
	CMP B #$0A	;Increment index
	BNE DIGIT	;Increment digit #
	JMP MP10	
		;Done all 10?
DATA:	DB 1,2,3,4,5,6,7,8,9	

chase the DAC chips and do the interfacing himself.

Digital-to-analog converters are used whenever the user has to generate a proportional analog voltage under software control. A typical DAC will have eight or more digital inputs, an analog voltage output, and two or more power and ground connections. Also, DACs are available in a number of sizes, ranging from a 16-pin integrated circuit to a module that is 3 inches wide, 4 inches long, and ½ inch thick.

The first DAC that we will use in an interface is the NE5018 (Fig. 3-10). It is an 8-bit DAC and can produce an analog voltage between 0–10 volts. This particular device has straight binary inputs; other DACs use BCD, 2's complement, and offset binary numbers. Since the NE5018 uses straight binary numbers, the eight inputs can assume all values between 00000000 and 11111111. Thus, there are 256 different states; the analog voltage produced will be between 0–10 volts. As you might guess, this converter cannot produce every possible analog voltage between zero and 10 V, simply because there are an infinite number of discreet analog voltages. Instead, it can produce 256 different analog voltages in this range. Therefore, the value 00000000 will produce 0 volt, the value 00000001 will produce 39 mV, the value 00000010 will produce 78 mv, and up to 11111111, which will produce 9.98 V. Using a potentiometer, this analog voltage can usually be adjusted up to exactly 10.00 volts. The *step* that is produced when the binary number is incremented by one is determined by dividing the range of the converter by the number of different input states (10/256, or $\cong$ 40 mV).

There are a number of other factors and specifications to be considered when using DACs but they will not be discussed here.[1,2] One important point to realize about many DACs is the fact that the analog output will start changing as soon as the digital inputs are changed. Thus, in many DAC interfaces, latches have to be used to latch data off of the data bus. A latch doesn't have to be used with the NE5018 converter, simply because a latch has already been built into the chip!

An interface based on the NE5018 is presented in Fig. 3-11. As you can see, the interface is very simple because the NE5018 contains an 8-bit latch. One of the easiest ways to experiment with a DAC and see the result is by wiring the voltage output of the DAC (V_{DAC}) to an oscilloscope. By doing this, you can see the analog voltage change as the digital inputs to the DAC are changed.

With the DAC interfaced to the microcomputer, some simple software can be used to generate a voltage ramp on the scope (Example 3-12). Regardless of the microprocessor present in your STD bus system, all that the software in Example 3-12 does is increment an 8-bit register and write it out to the digital-analog converter. Since we want to observe the ramp on an oscilloscope and not have it flicker, we want the microprocessor to execute this software as fast as possible. Therefore, there are no time-delay instructions in the software.

The "trick" in this software is the fact that when the register in the CPU chip contains FFH and is incremented by one, the register is incremented to 00H, which will cause the output of the DAC to go from 10 V to 0 V. Thus, the ramp is repeated again from the beginning. At this point, you may want to modify this software so that the ramp has a negative slope rather than a positive slope (Fig. 3-12). To do this, all you have to do is change the increment instruction to a decrement instruction.

How would a triangle waveform be produced? If you can come up with a solution and try it on your microcomputer, you may observe some "glitches" at the very top of the display (where the slope changes sign). The proper solution to this problem for the various processors is shown in Example 3-13.

The trick in this software is to make sure that when the register is incremented up to FFH and, then, is incremented to 00H, that the 00H is not output to the DAC. Instead, the 00H has to be decremented to FFH and, then, decremented again to FEH, and then output. When 00H is decremented to FFH, the FFH will not be output, because FFH was already output. If FFH was to be output a second

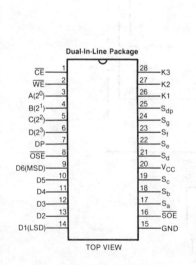

(A) Pin configuration.   (B) Truth tables.

FIGURE 3-8
The MM74C917 display controller chip. (Courtesy National Semiconductor Corp.)

105

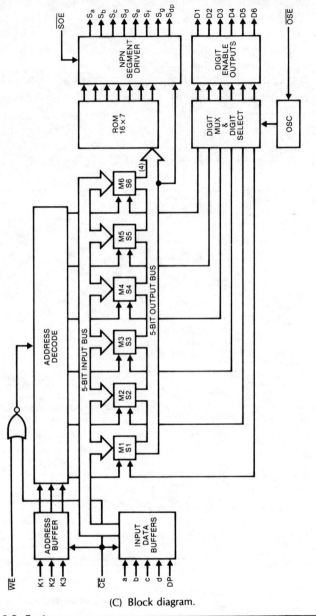

(C) Block diagram.

FIGURE 3-8. Cont.

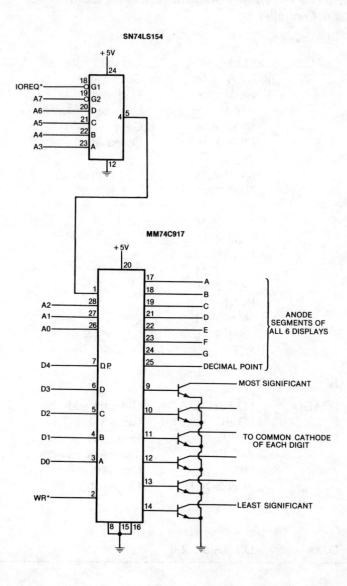

FIGURE 3-9

Interfacing an MM74C917 display controller to an STD bus microcomputer.

Example 3-10. Writing Six BCD Values to the MM74C917 Display Controller

1. Using the 8085.

LOADDC:	LXI H,DATA	;Get data address
	MOV A,M	;Get a value
	OUT 20H	;Output it
	INX H	;Increment address
	MOV A,M	;Get next value
	OUT 21H	;Output it
	INX H	
	MOV A,M	
	OUT 22H	
	INX H	
	MOV A,M	
	OUT 23H	
	INX H	
	MOV A,M	
	OUT 24H	
	INX H	
	MOV A,M	
	OUT 25H	
	RET	

2. Using the Z-80.

LOADDC	LD HL,DATA	;Get data address
	LD BC,0C20H	;Get a value
OUTIT	OUTI	;Output it
	INC C	;Increment address
	DEC B	;Get next value
	JR NZ,OUTIT	;Output it
	RET	
DATA	DB 9,5,6,3,2,8	

time, the time that the DAC would spend outputting 10 V would be twice as long as the time spent with any of the other 255 values.

Example 3-11. Writing Six BCD Values to the MM74C917 Display Controller

1. Using the 6502.

LOADDC	LDX #$06	;Set up an X
	LDY #$06	;and Y index
OUTIT	LDA DATA,X	;Get a value
	STA $C020,Y	;Output it
	DEX	;Decrement X
	DEY	;Decrement Y
	BNE OUTIT	;Done all 6?
	RTS	;Yes, return

2. Using the 6800.

LOADDC	LDX DATA	;Get a value
	STX $0020	
	LDX #$C020	
	STX $0022	
	LDA B #$06	
OUTIT	LDX $0020	
	LDA A 0,X	
	INX	
	STX $0020	
	LDX $0022	
	STA A 0,X	;Output it
	INX	
	STX $0022	
	DEC B	;Decrement B
	BNE OUTIT	;Done all 6?
	RTS	;Yes, return
DATA	DB 9,5,6,3,2,8	

DATA DISPLAYS

Although generating waveforms on an oscilloscope is interesting, there are not many applications that can use these low-frequency signals. There are far more applications that use DACs to *display data* on a display. To do this, information stored in memory has to be

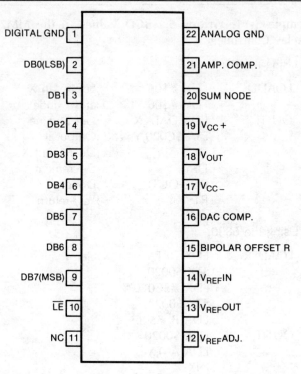

FIGURE 3-10
Pin configuration of the NE5018 8-bit digital-to-analog converter. *(Courtesy Signetic Corp.)*

output to the DAC. At the same time, the electron beam of the scope must move from left to right so that as new data are output, the electron beam goes to different X (time) and Y (displacement) coordinates. A simple example of this is shown in Example 3-14.

In Example 3-14, the first "I/O" instruction is used to generate a device-select pulse, which triggers the oscilloscope. Once triggered, the electron beam moves from left to right, regardless of what the microprocessor does next. The microcomputer then writes 256 8-bit data values to the DAC as quickly as possible. Once this is done, the entire process is repeated, so that an image appears on the scope face.

One problem with this type of display is the fact that the microcomputer really doesn't have much control over the electron beam once it is triggered. A better solution for a display involves the

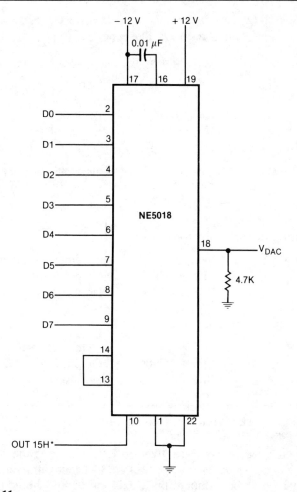

FIGURE 3-11
An STD bus interface using the NE5018 8-bit DAC.

use of two DACs—one for data (Y axis) and one for a position (X axis). A dual DAC interface is shown in Fig. 3-13. If you need specifics, just duplicate the circuitry shown in Fig. 3-11 twice, using different device-select pulses to latch the information into the DACs. With this type of interface, the oscilloscope is no longer triggered. Instead, the microcomputer has to output the X position to one DAC and the data to the other DAC. The software that does this (Example

Example 3-12. Software for Using a DAC to Generate a Positive Voltage Ramp on an Oscilloscope

1. Using the 8085 microprocessor:

RAMP:	INR A	;Increment register A
	OUT 00H	;Output it to DAC
	JMP RAMP	;Do it again

2. Using the Z-80 microprocessor:

RAMP	INC A	;Increment register A
	OUT (00),A	;Output it to DAC
	JP RAMP	;Do it again

3. Using the 6502 microprocessor:

RAMP	INX	;Increment register X
	STX $C000	;Output it to DAC
	JMP RAMP	;Do it again

4. Using the 6800 microprocessor:

RAMP	INC A	;Increment register X
	STA A $C000	;Output it to DAC
	BRA RAMP	;Do it again

3-15) is very similar to that given in the previous data-display software (Example 3-14).

By modifying the software given in Example 3-15, the dual DAC interface can also be used to control an X–Y (analog) plotter. As before, data would be output to the Y axis and a ramp would be output

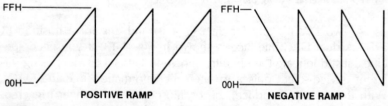

FIGURE 3-12

Positive and negative ramps generated by the microcomputer.

Example 3-13. Software for Generating a Triangular Wave on an Oscilloscope While Using a DAC

1. Using the 8085 microprocessor:

UP:	XRA A	;Clear register A
UP1:	OUT 15H	;Output to DAC
	INR A	;Increment A
	JNZ UP1	;Loop, not zero
	DCR A	;A = FFH
DOWN:	DCR A	;Decrement A
	OUT 15H	;Output value
	JNZ DOWN	;Loop until zero
	INR A	;A = 01H
	JMP UP1	;Ramp up

2. Using the Z-80 microprocessor:

UP	XOR A	;Clear register A
UP1	OUT (15H),A	;Output to DAC
	INC A	;Increment A
	JP NZ,UP1	;Loop, not zero
	DEC A	;A = FFH
DOWN	DEC A	;Decrement A
	OUT (15H),A	;Output value
	JP NZ,DOWN	;Loop until zero
	INC A	;A = 01H
	JP UP1	;Ramp up

3. Using the 6502 microprocessor:

UP	LDX #$00	;Load X with zero
UP1	STX $C015	;Output to DAC
	INX	;Increment X
	BNE UP1	;Loop until zero
	DEX	;X = FFH
DOWN	DEX	;Decrement X
	STX $C015	;Output to DAC
	BNE DOWN	;Loop until zero
	INX	;X = 01H
	JMP UP1	;Ramp up

Example 3-13. Cont.

4. Using the 6800 microprocessor:

UP	CLR A	;Load X with zero
UP1	STA A $C015	;Output to DAC
	INC A	;Increment X
	BNE UP1	;Loop until zero
	DEC A	;X = FFH
DOWN	DEC A	;Decrement X
	STA A $C015	;Output to DAC
	BNE DOWN	;Loop until zero
	INC A	;X = 01H
	BRA UP1	;Ramp up

to the X axis. However, the microprocessor must not be allowed to output information as fast as it can to the plotter. Since the plotter is

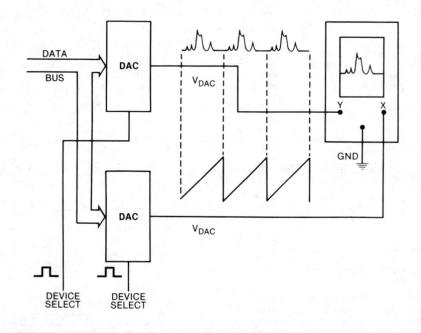

FIGURE 3-13

Using dual DACs to obtain an X-Y oscilloscope display.

Example 3-14. Displaying 256 Data Points on an Oscilloscope

1. Using the 8085:

DISPLY:	LXI H,DATA	;Set up an address
	MVI E,00H	;Set up a count
	OUT 40H	;Trigger the scope
LOOP	MOV A,M	;Get a data value
	OUT 15H	;Output to DAC
	INX H	;Increment address
	DCR E	;Decrement count
	JNZ LOOP	;Loop if not zero
	JMP DISPLY	;Do 256 points again

2. Using the 6502:

DISPLY	LDA #$00	;Save the address
	STA $40	;7000H on page
	LDA #$70	;zero.
	STA $41	
	LDY #$00	;Set up an index
FRAME	STA $C040	;Trigger the scope
LOOP	LDA ($40),Y	;Get a data value
	STA $C015	;Output to DAC
	INY	;Increment index
	BNE LOOP	;Loop if not zero
	JMP FRAME	;Do 256 points again

3. Using the Z-80:

DISPLY	LD HL,DATA	;Set up an address
	LD BC,0015H	;Set up a count
	OUT(40H),A	;Trigger the scope
	OTIR	;Output to DAC
	JP DISPLY	

4. Using the 6800:

DISPLY	LDX DATA	;Save the address
	CLR B	
	STA A $C040	;Trigger the scope

Example 3-14. Cont.

```
        LOOP      LDA A 0,X        ;Get a data value
                  STA A $C015      ;Output to DAC
                  INX              ;Increment index
                  DEC B
                  BNE LOOP         ;Loop if not zero
                  BRA DISPLY       ;Do 256 points again
```

an electromechanical device, we must worry about how fast the pen can traverse the plotter bed and the amount of momentum that the pen will have. Although this may sound complex, all you need to do is program the microcomputer to execute a short time delay, once the X and Y values have been output. This time delay will be the "worst case" time period that allows time enough for the pen to move from one side of the plotter to the other.

Another problem that has to be solved concerns the pen. The pen cannot be left pressed down on the paper as new data values are output to the plotter. Instead, once the pen is in position, it is dropped on the paper and a short time later it is lifted off of the paper. The next two values are output, etc. The reason that the pen must be lifted is due to the fact that the plotter will take an indirect path to get to its final position. *It will not move in a straight line.* This is due to the analog electronics contained in the plotter. Thus, your software has to both lift the pen up off the paper and, also, drop it down on the paper. Remember, the pen has to be left on the paper for a time period that is between 0.1 to 0.5 second in order for the ink to flow out of the pen and onto the paper. Thus, the only modifications required are some time-delay instructions and the adding of the pen up/down I/O instructions. Usually, all that this will require is two I/O instructions in the program and a flip-flop in the interface (Fig. 3-14).

OTHER DAC CONSIDERATIONS

In many applications, the data being processed has more resolution than one part in 256. Using 8-bit DACs, we are limited to this resolu-

Example 3-15. Controlling the X and Y Axes with DACs

1. Using the 8085 processor:

DDAC:	LXI H,DATA	;Set up an address
	MVI E,00H	;Set up a count
LOOP:	MOV A,E	;Get the count
	OUT 16H	;Output to X axis
	MOV A,M	;Get the data
	OUT 15H	;Output to Y axis
	INX H	;Increment address
	INR E	;Increment count
	JNZ LOOP	;Loop if not zero
	JMP DDAC	;Do 256 points again

2. Using the 6502 processor:

DDAC	LDA #00	;Save the address
	STA $40	;7000H on page
	LDA #$70	;zero.
	STA $41	
	LDY #$00	;Set up an index
LOOP	STY $C016	;Index to X DAC
	LDA ($40),Y	;Get data
	STA $C015	;Data to Y DAC
	INY	;Increment the index
	JMP LOOP	;Do it again

3. Using the Z-80 processor:

DDAC	LD HL,DATA	;Set up an address
	LD E,00H	;Set up a count
LOOP	LD A,E	;Get the count
	OUT (16H),A	;Output to X axis
	LD A,(HL)	;Get the data
	OUT (15H),A	;Output to Y axis
	INC HL	;Increment address
	INC E	;Increment count
	JP NZ,LOOP	;Loop if not zero
	JP DDAC	;Do 256 points again

Example 3-15. Cont.

4. Using the 6800 processor:

DDAC	LDX DATA	;Save the address
	CLR B	;Set up an index
LOOP	STA B $C016	;Index to X DAC
	LDA A 0,X	;Get data
	STA A $C015	;Data to Y DAC
	INX	
	INC B	
	BNE LOOP	
	JMP DDAC	;Do it again

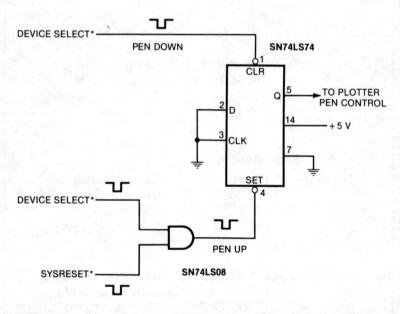

FIGURE 3-14
Pen control logic in an X-Y plotter interface.

tion. If this is a problem, you can use 10- or 12-bit DACs. However, the interface and the software become more complex.

In the interface, only 8 bits of information can be transferred on the data bus at a time, so if a 12-bit DAC is used, 8 bits would have to be output to it, followed by the remaining 4 bits (or *vice versa*). Regardless of the order in which the information is output, you have to ensure that the entire new 12-bit word is present on the *DAC's inputs* at the *same time*. Large glitches will occur if this is not done. This is called double-buffering and has actually been built into many DACs, along with the latches. The point is, the DAC now requires three device-select strobes, one for the 8-bit transfer, one for the 4-bit transfer, and one to load the DAC (not the latches) with all 12 bits of information.

The software is also more complex. The data values being output must now be stored in two memory locations rather than one, and some of the bits in one of the memory locations will be wasted. Thus, to store 1000 12-bit data values, you will need 2000 memory locations. The software will also take longer to execute. (This is particularly important for oscilloscope displays.) If you are interested in an oscilloscope display, you may really have to "tweak" your software and use some tricks to get as much speed as possible out of your microcomputer. As displays start to contain 2000 or more data points, the flicker becomes more noticeable.

The only solution to this problem is to either design and built a display-controller board or find a commercially available board that meets your needs. In general, these boards will contain R/W memory chips which are separate and distinct from the memory chips that compose the microcomputer's main memory. Information can be clocked out of this buffer memory every 500 nsec or so, which is far faster than the microprocessor can read a value from memory and output it to a DAC.

I/O CHIPS

There are a number of complex peripheral interface chips that contain both output ports and input ports (see Chapter 4). Some of these chips are general-purpose I/O chips that can be used to interface devices such as DACs to the microprocessor. Other I/O chips have very specific functions, such as serial communications, floppy disk control, video display (nonDAC) control, and direct-memory access control.

MEMORY-MAPPED OUTPUT PORTS

As you might guess, there is really very little difference between an accumulator output port and a memory-mapped output port. The only differences in the interface are the use of the MEMRQ* signal instead of the IORQ* signal and a 16-bit address that will be used instead of an 8-bit address.

The differences in the software produce an interesting contrast. For the 6800 and 6502, no changes are required in the software, other than changing a single memory address. Remember, the 6502 and 6800 can only communicate with memory. Thus, to communicate with peripherals, the peripherals have to look like memory locations. In order for the processors to meet the STD bus definition, they have to decode a block of 256 addresses on the CPU card, and, then, generate the IORQ* signal when one of these 256 addresses is present on the address bus. At the same time, the processor card logic has to prevent the MEMRQ* signal from going to the bus. Therefore, in the software examples for the 6800 and 6502, the I/O address C000 might be changed to 80FE or 5060.

Since the 8085 and Z-80 can use both accumulator I/O and memory-mapped I/O, different instructions have to be used to communicate with memory-mapped I/O peripherals. As was the case with the 6800 and 6502, these processors have to use instructions that write information into memory instead of using OUT-type instructions. For the 8085, we might use MOV M,A, STAX D, STAX B, or MOV E,M. On the Z-80, you could use LD (HL),A, LD (DE),A, LD (BC),A, LD (IX),C, or LD (HL),54H.

In Examples 3-10 and 3-11, the 8085 had the longest program, simply because an accumulator I/O interface was used, which meant that accumulator I/O instructions had to be used in the software examples. By changing the interface to memory-mapped I/O, the software becomes a lot simpler (Example 3-16). *Remember, the only difference in the interface is that a 16-bit address and MEMRQ* are used to generate the device-select pulse, rather than an 8-bit address and IORQ*.* So, the interfaces used in this chapter will not be redesigned, using memory-mapped I/O.

In general, in order to convert the 8085 and Z-80 software from accumulator I/O to memory-mapped I/O, the OUT instructions would be changed to the 3-byte STA instruction (8085) or the 3-byte LD (nn),A instruction (Z-80). Thus, it is relatively easy to switch between accumulator I/O and memory-mapped I/O.

Example 3-16. Writing 6 Data Values to a Memory-Mapped MM74C917 Display Controller

1. Using the 8085:

LOADDC:	LXI H,DATA	;Data address
	LXI D,FFF0H	;Display address
	MVI C,06H	;Count
OUTIT:	MOV A,M	;Get a data value
	STAX D	;Output it
	INX H	;Increment data address
	INX D	;Increment display
	DCR C	;Decrement count
	JNZ OUTIT	;Loop if not zero
	RET	

2. Using the Z-80:

LOADDC	LD HL,DATA	;Data address
	LD DE,FFF0H	;Display address
	LD BC,0006H	;Count
OUTIT	LDIR	;Output a data value
	RET	
DATA	DB 9,5,6,3,2,8	

CONCLUSION

In this chapter, you have seen how an output port is designed, along with the software that can be used to access this type of I/O device. Regardless of what the output port is used for, it must contain some type of *latch*. These latches, along with the appropriate drivers, can be used to control LEDs, solid-state relays, 7-segment displays, and digital-to-analog converters. There are also a large number of peripherals that were not discussed which use output ports for the control and the transfer of information. These include stepper motors, floppy disks, serial communications ports, security systems, audio cassette interfaces, bank-select systems for large amounts of "main" memory, and parallel printers.

REFERENCES

1. Titus, J.A., Larsen, D.G., and Titus, C.A., *Microcomputer–Analog Converter Software and Hardware Interfacing,* Howard W. Sams & Co., Inc., Indianapolis, IN, 1978.
2. *Analog-Digital Conversion Handbook,* D. Sheingold, ed., Analog Devices, Norwood, MA, 1972.

Input Ports
Chapter 4

In Chapter 3, we saw how output ports are used by the microcomputer to output information to peripherals. Once this information is output, it can be displayed, stored on a floppy disk, or printed on a line printer. In this chapter, we will examine the requirements for an input port interface—the timing involved and the chips that are used.

Unlike output ports that must be able to accept and hold data at a specific time, and which may be continuously connected to the data bus, input ports must be able to "disconnect" themselves from the data bus whenever they are not in use. The input ports must pass logical information (logic 1s or 0s) and they must also be configured so that they do not interfere with the use of the data bus during the time when they are not selected.

Simple gates cannot be used to gate data onto the data-bus lines since, depending on the type of gate used, their "unselected" output will be either a logic 1 or a logic 0. This is shown in Fig. 4-1. Note that even when none of the gates are selected or enabled, the outputs of the gates generate different logic levels, as noted by the quoted logic levels. These levels all compete for the use of the bus, probably leading to one or more burned out chips. Remember, the CPU chip uses the data bus to transfer information between itself and memory and peripheral devices. Thus, an input port must only put information on the data bus when the microprocessor requests it. At all other times, the input port should not use, or affect the information that is on, the data bus.

Special intergrated circuits are available that have *three-state outputs*, which greatly simplify the design of input ports. A typical

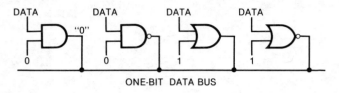

FIGURE 4-1
Attempting to use standard gates on a data bus.

three-state device is the SN74125 bus buffer (Fig. 4-2). This chip contains four buffers (logic 1 in, logic 1 out). It has an additional control line for each buffer, shown in Fig. 4-2 as connected to one of the angular sides of the buffer symbol.

The buffer will pass logic 1s and 0s from its input to its output when it is *enabled* but, unlike a simple gate, the output of the buffer appears to be electrically disconnected from whatever it is wired to when the buffer is disabled. In three-state devices, the "third state" is called the Hi-Z, or high-impedance, state. When one of these three-state devices is used in an input port, the information from the peripheral is wired to the inputs of the buffers, and the three-state outputs of the buffers are wired to the data bus.

DESIGNING INPUT PORTS

To demonstrate how these devices are used, let's examine one line in the data bus and see how four 1-bit peripherals are wired to this

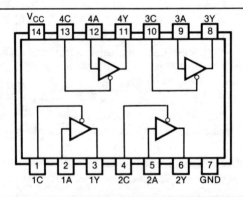

FIGURE 4-2
Pin configuration for the SN74125 quadruple bus buffer with three-state outputs.

single data-bus line. The circuit for this is shown in Fig. 4-3. Of course, in many applications, we would have to use all 8 bits of the data bus. From Fig. 4-3, you can see that when one of the control inputs is a logic 0, the data present on the input of the buffer will pass through the buffer and out, onto the single data-bus line. For the SN74125, the enable inputs are active in the logic 0 state, thus the use of the asterisks next to the "device names." If the enable input is in the logic 1 state, the output of the buffer will be in the Hi-Z state. Based on this description, a truth table can be created that will describe the operation of these four 1-bit devices (Table 4-1).

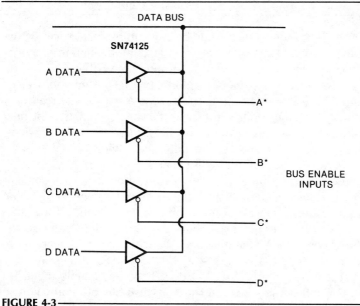

FIGURE 4-3
Using three-state devices as four 1-bit devices.

From this truth table, you can see that if none of the buffers are enabled, we cannot tell what the state of the bus will be, because we don't know whether the microprocessor is reading from memory, writing to memory, communicating with an output port, or even doing some internal operation. However, whenever one of the buffers is enabled (a logic 0 is present on its control input), whatever is present on the input of the buffer will be passed onto the data bus. As you might expect, the condition where two or more buffers are enabled is not allowed, because one device might be trying to place a logic 0 on

Table 4-1. Truth Table for Four 1-Bit Three-State Devices

Enable				Bus Content
D	C	B	A	
1	1	1	1	Indeterminate. (All devices Hi-Z.)
1	1	1	0	A data
1	1	0	1	B data
1	0	1	1	C data
0	1	1	1	D data
0	0	0	0	Not allowed. (Any state where two or more enable inputs are in the logic 0 state.)

the bus, and the other device might be trying to place a logic 1 on the bus. When you think about it, you can't read information from two or more memory locations at exactly the same time either.

All of the devices that transfer information to the microprocessor chip must have three-state outputs. Thus, R/W memories, read-only memories, and input ports must have three-state outputs. Not only must three-state devices be used for all of these input-like devices, but the designer must ensure that only one device is enabled at a time, otherwise a *bus conflict* will occur.

Because so many three-state devices are used in microcomputer interfaces, semiconductor manufacturers have designed chips that contain 8 buffers, with all of the enable inputs connected together. Thus, a state-of-the-art three-state buffer will have 8 inputs, 8 outputs, and 1 or, possibly, 2 control inputs. Typical chips that meet these specifications are the SN74LS244 and the SN74LS373 (Fig. 4-4).

As you can see, the SN74LS244 does have two enable inputs. In most input port designs, we *want* to transfer 8 bits of information so these two inputs would be wired together and would be driven by a combination of a device address (RD*) and either IORQ* (accumulator I/O) or MEMRQ* (memory-mapped I/O). A typical input port interface is shown in Fig. 4-5.

In Fig. 4-5, you can see that the "peripheral" consists of a group of eight switches. The information from these switches, between 0 and 255_{10}, will be input into the microprocessor whenever the microprocessor reads from the accumulator I/O device E5H. Note that we have assumed that the address decoder is not enabled with either IORQ* or RD* so these signals have to be gated together, and the resulting signal is gated with the logic 0 output of the address de-

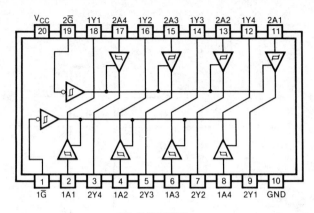

(A) SN74LS244.

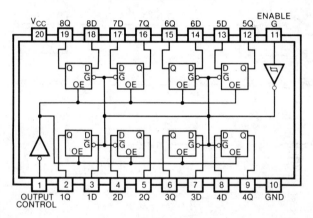

(B) SN74LS373.

FIGURE 4-4

Pin configurations for two three-state buffers.

coder. Therefore, whenever the microprocessor inputs information from device E5, a logic 0 pulse will occur on pins 1 and 19 of the SN74LS244 chip. The three-state devices will be enabled and the state of the switches will be gated into either the accumulator or the A register (8085 and 6502), either the A or B accumulators (6800), or into any general-purpose register (Z-80; A-E, H or L).

How long will the switch information actually be on the data bus? This really depends on the type of processor used in the STD bus system and its cycle time. Referring to the timing of the Z-80 again

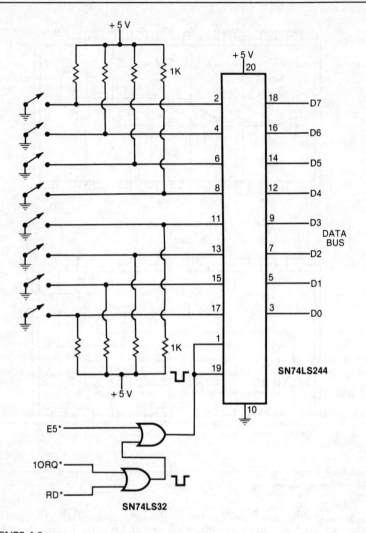

FIGURE 4-5

An 8-bit input port based on the SN74LS244 8-bit three-state buffer.

(Fig. 4-6 and Table 3-1), you can see that the duration of the RD* and IORQ* signals is 375 nsec for a 4-MHz Z-80. The Read times for all of the STD bus processors are summarized in Table 4-2.

As you can see, the information from the switches is only on the data bus for a very short period of time, and *only when the processor*

Table 4-2. Typical Read Times for STD Bus Processors

CPU Chip	Cycle Time	Read Time
8085	320 nsec	400 nsec
Z-80	250 nsec	375 nsec
6800	1000 nsec	450 nsec
6502	1000 nsec	430 nsec
6809	1000 nsec	450 nsec
8088	200 nsec	325 nsec

requests the information (by taking IORQ* and RD* to the logic 0 state). Once the information has been input, the outputs of the SN74LS244 go back into the high-impedance state so that the microprocessor can use the data bus to communicate with memory or output devices.

The software for this input-port interface is trivial. It consists of a single two- or three-byte instruction, regardless of the processor. For the 8085-type processors (8085, 8088, and Z-80), this would be some type of IN instruction and, for the 6800-type processors (6800, 6502, and 6809), this would be some type of LDA instruction.

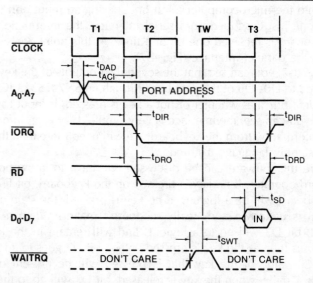

FIGURE 4-6
READ timing for a Z-80 CPU card. *(Courtesy Mostek Corp.)*

In the interface shown in Fig. 4-5, a set of eight switches is used to generate information that the microcomputer can input. Even though this may not seem like a very practical example, there are a number of situations where simple switches are interfaced to microcomputers. The switches could be limit switches on a numerically controlled machine and the microcomputer is used to control the machine. The switches could also be thermostatic switches that are closed when the temperature being measured is below a set temperature and open when the temperature is above the set temperature. These switches might represent the state of doors and windows (open or closed), or the state of valves, or they might actually represent a binary data value that the microcomputer should use in a mathematical calculation.

AN ASCII KEYBOARD INTERFACE

One of the most common peripheral devices in microcomputer systems is the ASCII keyboard. This device can be interfaced using two entirely different techniques: (1) parallel interfacing, or (2) serial interfacing. In a parallel interface, all seven or eight bits of information are input into the microcomputer all at once, using an input port similar to the one just described. In a serial interface, the information from the keyboard is received one bit at a time. At this point, we will just discuss the parallel interface (Fig. 4-7).

In Fig. 4-7, you can see that the seven data outputs of the keyboard are wired to D0–D6 of the data bus (through an SN74LS373 IC). One bit, which indicates whether or not a key is pressed, is input using bit D7. As you can see, the interface is very simple. However, in order to input information from the keyboard, we will need more than three or four assembly language instructions.

Before the software can be discussed, we have to know how the keyboard operates. If no key is pressed on the keyboard, bit D7 from the keyboard will be a logic 0. If no key is pressed, the state of lines D0–D6 is unknown and, really, it doesn't matter. When a key is pressed, bit D7 will go to a logic 1, and will remain in the logic 1 state for as long as the key is pressed. When bit D7 goes to a logic 1, the ASCII value of the key that is pressed will be present on bits D0–D6. Finally, when the key is released, bit D7 will go to the logic 0 state. If one key is pressed and, then, another key is pressed, the ASCII value generated by the keyboard will represent the first key that is pressed. The second key will be ignored until the user releases both keys and, then, presses the second key again.

Unfortunately, the keys in this particular ASCII keyboard are not perfect. Thus, when a key is pressed, the key "contacts" bounce, so there are no clean transitions between the logic 0 and logic 1 state. This is shown in Fig. 4-8. The transitions or "bounces" between these two logic states can be cleaned up with both hardware and software techniques. A hardware approach is shown in Fig. 4-9, where the keyboard strobe signal (bit D7) has to be clocked through two flip-flops. The software that takes key bounce into consideration is given in Examples 4-1 and 4-2.

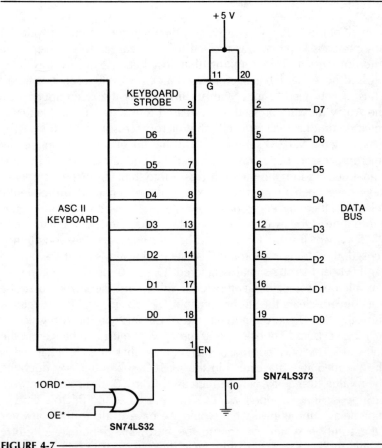

FIGURE 4-7
A simple 8-bit ASCII keyboard interface.

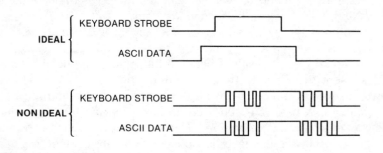

FIGURE 4-8
Ideal and nonideal keyboard signals.

In all of the keyboard subroutines, the microprocessor first waits for any previous key that was pressed to be released. This prevents the microprocessor from thinking that just because a key has been pressed once that it has been pressed a number of consecutive times. Thus, if you need to input the ASCII value for the A key four times, the A key must be pressed and released four times. Once the key is released (if a key was pressed), the microprocessor waits 10 msec by executing the MS10D subroutine. This has the effect of locking up the microprocessor for 10 msec, so that the bouncing keyboard strobe signal doesn't fool the microprocessor into thinking that a number of keys have been pressed. If no key was pressed (and subsequently released), the micrprocessor simply wastes 10 msec of its time in this subroutine. One way or another, the microprocessor makes it to PRESS where it waits for a key to be pressed. Once this condition is sensed, the 10-msec subroutine is executed again, so that the bouncing keyboard strobe signal is ignored. Once 10 msec have gone by, the microprocessor inputs the ASCII value (and the keyboard strobe logic level), clears the strobe signal at D7 to zero, and, then, returns from the KYBD subroutine with the ASCII value in the A register.

If the keyboard bounces for 20 msec, what must be changed in the software? The only portion of the software that must be changed is the "immediate data bytes" in the MS10D subroutine. By doubling their value, the delay will be increased from 10 msec to 20 msec.

Suppose that we don't want to waste a lot of the micro's time in time-delay subroutines, what would we have to do to the software? One solution would be to use the hardware debounce circuitry shown in Fig. 4-9. This would mean that the MS10D subroutine and all references to it could be deleted from the KYBD subroutines.

Example 4-1. Debouncing and Reading an ASCII Value From a Keyboard

1. Using an 8085 microprocessor:

KYBD	IN 0EH	;Get flag & data
	RLC	;Flag into carry
	JC KYBD	;Loop if a 1
	CALL MS10D	;10-msec delay subroutine
PRESS	IN 0EH	;Get flag & data
	RLC	;Flag into carry
	JNC PRESS	;Wait, flag is 0
	CALL MS10D	;Pressed, wait 10 msec
	IN 0EH	;Get flag & data
	ANI 7FH	;Flag = 0
	RET	
MS10D	PUSH D	;Save D & E
	LXI D,0514H	;Load count
MS10D1	DCX D	;Decrement count
	MOV A,D	;Get MSBs of count
	ORA E	;OR with LSBs
	JNZ MS10D1	;Not zero, loop
	POP D	;Restore D & E
	RET	

2. Using a Z-80 microprocessor:

KYBD	IN A,(0EH)	;Get flag & data
	RLCA	;Flag into carry
	JR C,KYBD	;Loop if a 1
	CALL MS10D	;10-msec delay subroutine
PRESS	IN A,(0EH)	;Get flag & data
	RLCA	;Flag into carry
	JR NC,PRESS	;Wait, flag is 0
	CALL MS10D	;Pressed, wait 10 msec
	IN A,(0EH)	;Get flag & data
	AND 7FH	;Flag = 0
	RET	

Example 4-1. Cont.

MS10D	PUSH DE	;Save D & E
	LD DE,0533H	;Load count
MS10D1	DEC DE	;Decrement count
	LD A,D	;Get MSBs of count
	OR A,E	;OR with LSBs
	JR NZ,MS10D1	;Not zero, loop
	POP DE	;Restore D & E
	RET	

FLAGS

In all of the previous examples, except the ASCII keyboard interface, we have assumed that there is no synchronization required between the peripheral and the microcomputer. Thus, the microcomputer either inputs or outputs information when *it* wants to. We have also assumed that the peripheral either can accept the information (output ports) or has information ready for us (input ports). In many cases, this is not a valid assumption. With the ASCII keyboard software, the microcomputer has to be programmed to wait for a key to be pressed before an ASCII value can be input from the keyboard. Thus, the computer is synchronized with the keyboard via the *keyboard strobe flag*.

In another example, let's assume that we want to write information out to a floppy disk. The software has to first output the sector and track information to the floppy disk and, then, the head will start to move to the appropriate track. The sector numbers will then be read by the interface until the proper sector has been found. Thus, once the sector and track information are output to the disk, the microcomputer would not immediately start to output the information that is to be written onto the disk. Instead, the microcomputer must wait for a flag (a logic level), generated by the floppy disk interface, to tell the microcomputer that the proper track and sector have been found and that it is time to write the information onto the disk. Depending on the track and sector required, it may take the disk 40–50 msec to get to the right place, before data can be written onto the disk. Thus, flags are used to synchronize the microcomputer to I/O

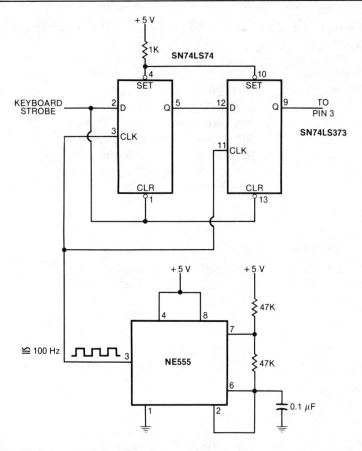

FIGURE 4-9
Debouncing the keyboard strobe signal with clocked logic.

devices, and flags are usually just logic levels. These logic levels will indicate whether or not the I/O devices are busy or not busy, ready or not ready, converting or not converting, etc. Thus, flags indicate the status of the I/O device. They are often called *status flags*.

In Fig. 4-7, the ASCII keyboard that was interfaced to the STD bus microcomputer only generated a 7-bit ASCII value. In some keyboards, 8 bits of information are generated; a 7-bit ASCII value and 1 bit of parity. In addition, the keyboard strobe signal would still be generated by the keyboard. How would this 8-bit keyboard be interfaced to the microcomputer?

Example 4-2. Debouncing and Reading an ASCII Value From a Keyboard.

1. Using a 6502 microprocessor:

KYBD	BIT $C00E	;Test flag bit
	BMI KYBD	;Loop if a 1
	JSR MS10D	;Wait 10 msec
PRESS	BIT $C00E	;Test flag bit
	BPL PRESS	;Loop if a 0
	JSR MS10D	;Pressed, so wait
	LDA $C00E	;Get flag and data
	AND #$7F	;Flag = 0
	RTS	
MS10D	LDY #$0A	;Load Y with 10
	LDX #$00	;Load X with 0
MS10D1	DEX	;Decrement X
	BNE MS10D1	;Loop if not 0
	DEY	;Decrement Y
	BNE MS10D1	;Loop if not 0
	RTS	

2. Using a 6800 microprocessor:

KYBD	BIT $C00E	;Test flag bit
	BMI KYBD	;Loop if a 1
	JSR MS10D	;Wait 10 msec
PRESS	BIT $C00E	;Test flag bit
	BPL PRESS	;Loop if a 0
	JSR MS10D	;Pressed, so wait
	LDA A $C00E	;Get flag and data
	AND A #$7F	;Flag = 0
	RTS	
MS10D	LDX #$04E2	;Load X with 1250
MS10D1	DEC X	;Decrement X
	BNE MS10D1	;Loop if not 0
	RTS	

The interface shown in Fig. 4-10 could be used. The important point here is that two separate and distinct input ports have to be used. Input port OF is used to input one bit of information (the status of the keyboard). If this bit is a logic one, the 8-bit ASCII value can be input. The I/O software for this interface is very similar to the software in Examples 4-1 and 4-2, except that a different I/O address is used for both the status flag and the ASCII value.

If you have a number of I/O devices that generate flags (disk interfaces, analog-to-digital converters, keyboards, magnetic sensors, etc.), you could dedicate all eight bits of input port OF to status flags and, then, use additional input ports for data. If you want, you can also build a number of 1-bit input ports—one flag per port. This is quite all right, but it is a little wasteful in terms of three-state ICs and the I/O address space.

ANOTHER KEYBOARD INTERFACE

There are a large number of different keyboards that you can interface to your microcomputer. Some keyboards produce a strobe signal for as long as a key is pressed, while some keyboards produce a single 500 nsec to 2 μsec pulse which indicates that a key is pressed (Fig. 4-11). Unfortunately, we cannot use either the previous interface

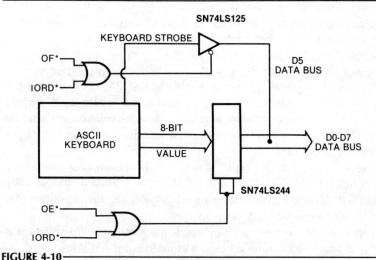

FIGURE 4-10
Using two input ports in an ASCII keyboard interface.

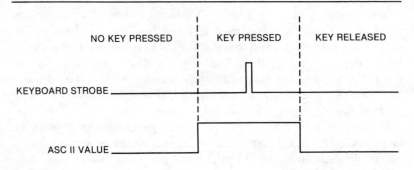

FIGURE 4-11
The timing relationship between the keyboard strobe pulse and the ASCII value.

or the software examples with this type of keyboard, simply because the pulse is so short; the microcomputer will not be able to detect it.

However, an interface that can be used with this type of keyboard is shown in Fig. 4-12. As you would expect, we still have to use an 8-bit input port in order to input the ASCII value for the key that is pressed. When a key is pressed, the short keyboard strobe pulse clocks the D-type flip-flop, and the Q output goes from a logic 0 to a logic 1. This is detected in our software by testing the state of bit D0 from input port OF. When this is a logic 1, the microcomputer inputs the ASCII value by reading from input port OE. The device-select pulse that enables the SN74LS244 also clears the flip-flop so that the Q output goes back to the logic 0 state. Only when the key that is pressed is released, and another key is pressed, will the flip-flop be clocked and allow the Q output to go back to the logic 1 state.

Note that the IN 0E* device-select pulse has been used to perform two tasks. This pulse is used to enable the SN74LS244, so that the keyboard data is gated onto the data bus, and it is also used to clear the flip-flop. If you wanted to, a separate device-select pulse could be used to enable the three-state device and another pulse could be used to clear the flip-flop.

The software that can be used with this keyboard and interface is listed in Example 4-3. There are a number of keyboards that already contain hardware that debounces the keys and we have assumed that this type of keyboard has been used.

The point of studying the keyboard interfaces and I/O software is not that you will become an expert at interfacing ASCII keyboards to an STD bus microcomputer. Instead, you should realize that three-

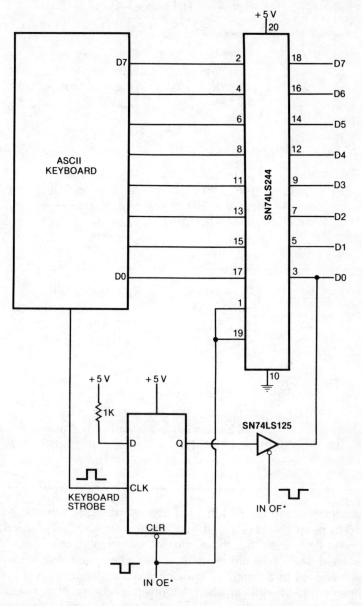

FIGURE 4-12
Capturing the keyboard strobe signal.

Example 4-3. Software for a Debounced ASCII Keyboard.

1. Using an 8085 processor:

```
KYBD:     IN 0FH        ;Get flag value
          ANI 01H       ;Save just the flag
          JZ KYBD       ;Flag is 0, loop
          IN 0EH        ;Get data
          RET
```

2. Using a 6502 processor:

```
KYBD      LDA $C00F     ;Get flag value
          AND #$01      ;Mask 7 bits out
          BEQ KYBD      ;Flag is 0, loop
          LDA $C00E     ;Get data
          RTS
```

3. Using a Z-80 processor:

```
KYBD      IN A,(0FH)    ;Get flag value
          AND 01H       ;Save just the flag
          JR Z,KYBD     ;Flag is 0, loop
          IN A,(0EH)    ;Get data
          RET
```

4. Using a 6800 processor:

```
KYBD      LDA A $C00F   ;Get flag value
          AND A #$01    ;Mask 7 bits out
          BEQ KYBD      ;Flag is 0, loop
          LDA A $C00E   ;Get data
          RTS
```

state devices must be used in input port interfaces, and that flags are used by peripherals to signal the microcomputer when they have data or need data. Even though we did not use any examples of output ports and flags, there are many examples where these two ideas are combined, as in a floppy disk interface where information is being written out to the disk, or where information is being written out to a line printer or crt. In the remainder of this chapter, we will see some applications for input ports and flags.

AN ANALOG-TO-DIGITAL CONVERTER INTERFACE

In Chapter 3, you saw how digital-to-analog converters are used by the STD bus microcomputer to generate analog waveforms. There are also analog-to-digital converters (ADCs) that are used to convert analog signals to digital values so that the microprocessor can process the "analog" information. Like DACs, ADCs come in a variety of shapes and sizes, ranging from an 18-pin DIP to a large multichip module. These devices may require +5-V and ±15-V power supplies, and may have 8, 10, 12, or even 16-bit outputs. The size of the analog voltage swing may also vary, with typical full-scale input values being between 10 mV and 10 V.

If you have a 10-bit, 0–5 V ADC, what is the smallest analog voltage that can be resolved? The smallest voltage that the microprocessor can measure would be 5 V/1024, or 4.88 mV.

Unlike DACs, ADCs require a large amount of time in order to convert the analog input to the proper digital output. Typically, ADCs require 10 μsec to 16 msec to perform this conversion. There are ADCs that are a lot faster than this that can digitize a 100-MHz analog signal, but their application in microprocessor-based systems is very limited, just because the computer is so much slower than the converter.

Another difference between ADCs and DACs is the fact that we have to tell the ADC when to start the conversion. With DACs, the conversion begins as soon as a new digital value is output. In 10–100 nsec, the analog output changes to reflect the change on the digital inputs. This means that the microcomputer has to strobe the ADC, start the converter, and, then, it has to wait for the BUSY or DONE flag (from the converter) to change states, which indicates that the analog signal has been completely digitized. Once the microcomputer senses this, the data value can be input from the ADC. This timing is shown in Fig. 4-13 and a typical 10-bit ADC interface is shown in Fig. 4-14.

Analog Devices AD571 is a 10-bit ADC, so two input ports are needed in the interface. One input port is used to input the 8 least-significant bits (LSBs) of the 10-bit value, and the other input port is used to input the two most-significant bits (MSBs) and the $\overline{\text{DATA READY}}$ (DATA READY*) flag. The only portion of this circuitry that you probably won't be familiar with is the part where we use NAND gates with the $\overline{\text{DATA READY}}$ flag and the CONVERT input.

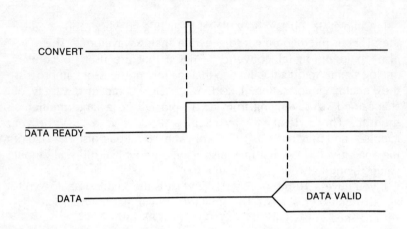

FIGURE 4-13
Typical timing for an analog-to-digital converter.

The CONVERT input of the AD571 requires a pulse at least 2 μsec in duration to start the converter. Using an STD bus microcomputer, the duration of the RD* or WR* signals is only a few hundred nanoseconds. Therefore, we need some circuitry to "stretch" one of these pulses. The OUT 36* pulse causes the CONVERT input to go to a logic 1 and, then, from 15 to 40 μsec later, when the analog-to-digital conversion has been performed, the logic 0 signal on the DATA READY output causes the logic level on the CONVERT input to go back down to a logic 0.

In order to initiate a conversion and input the 10-bit digital value, the microcomputer has to pulse the OUT 36* line and monitor bit D7 of input port 37H. When it is a logic 0, data can be input from input ports 36H and 37H. The software that does this is shown in Example 4-4. Regardless of the processor used, the software is basically the same. The only real difference is that the 8085, Z-80, and 6800 subroutines store the ADC data in registers, while the 6502 stores the ADC data in memory.

Ordinarily, these subroutines would be called by a "main program" where the main program either executes some type of time-delay software or starts a timer interfaced to the microcomputer, so

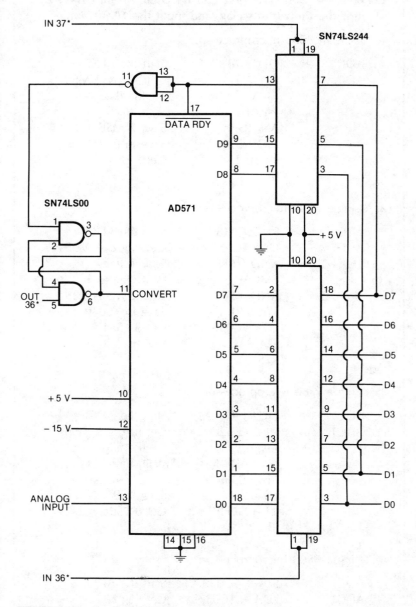

FIGURE 4-14
Interfacing a 10-bit analog-to-digital converter to the STD bus.

Example 4-4. Software That Can Be Used To Start the ADC, Monitor the Converter's Flag, and Input the 10-Bit Value.

1. Using an 8085 microprocessor:

ADC:	OUT 36H	;Start the ADC
ADC1:	IN 37H	;Get flag & 2 MSBs
	ANA A	;Set the flags
	JM ADC1	;Wait, flag = 1
	ANI 03H	;Save 2 MSBs
	MOV B,A	;Save in B
	IN 36H	;Get 8 LSBs
	RET	

2. Using a 6502 microprocessor:

ADC	STA $C036	;Start the ADC
ADC1	LDA $C037	;Get flag & 2 MSBs
	AND #$FF	;Set the flags
	BMI ADC1	;Wait, flag = 1
	AND #$03	;Save 2 MSBs
	STA $40	;Save in $0040
	LDA $C036	;Get 8 LSBs
	STA $41	;Save in $0041
	RTS	

3. Using a Z-80 microprocessor:

ADC	OUT (36H),A	;Start the ADC
ADC1	IN A,(37H)	;Get flag & 2 MSBs
	AND A	;Set the flags
	JP M,ADC1	;Wait, flag = 1
	AND 03H	;Save 2 MSBs
	LD B,A	;Save in B
	IN A,(36H)	;Get 8 LSBs
	RET	

4. Using a 6800 microprocessor:

ADC	STA A $C036	;Start the ADC
ADC1	LDA A $C037	;Get flag & 2 MSBs
	AND A #$FF	;Set the flags

Example 4-4 Cont.

```
          BMI ADC1        ;Wait, flag = 1
          AND A #$03      ;Save 2 MSBs
          TAB
          LDA A $C036     ;Get 8 LSBs
          RTS
```

that every 2 milliseconds, 5 seconds, or 2.374 hours, the analog signal is digitized and stored in memory. While the microcomputer is waiting for this time period to elapse, it might be displaying on a crt the last 500 data points that were acquired, or it might be executing a BASIC program.

There are a lot of details concerning the use of ADCs that we have not discussed. If you are interested in using ADCs with your STD bus microcomputer, you should refer to the references at the end of Chapter 3. Like the DACs discussed in Chapter 3, you can interface ADCs to your microcomputer using the circuitry that we have described, or you can buy a board that contains one or more ADCs, along with a sample-and-hold and an analog multiplexer. The sample-and-hold is an analog device that is used to memorize the analog signal present on one of the analog inputs to the card. This device is used so that (while the ADC is converting) the analog input to the ADC (the sample-and-hold output) doesn't change. If the analog signal does change during the conversion, the digital result will be in error.

If you have 10 analog signals that you want to digitize, you might interface 10 ADCs to the microcomputer. However, another way to solve this problem (and cut down on the amount of hardware required) is to use an analog multiplexer. By writing a digital word out to the multiplexer, the user can select which one of the analog inputs (usually there are 8 or 16 analog inputs) will be connected to the sample-and-hold input. Thus, by selecting a "channel" and performing a conversion, the user can digitize the analog signals on all 8 or 16 channels. The only disadvantage of this technique is the fact that if you have to digitize signals from all 16 channels, you can only digitize the signal on one channel at a time. Thus, using a multiplexer may mean that you can't digitize all the channels as quickly as you would like.

A SIMPLE LOGIC TESTER

In this chapter and in Chapter 3, the discussions were based on either input devices or output devices (with or without flags). Of course, it is possible to have peripherals that both need data (output) and can generate data (input) for the STD bus microcomputer. These I/O devices may also need flags in order to assist in the transfer of data.

One of the simplest interfaces that uses both input and output ports is a TTL logic device tester. Although this may not be a particularly practical or cost-effective way of testing ICs, it does demonstrate the interaction of both types of I/O ports.

From Figs. 2-1 and 2-7, you can see that the SN7400 NAND gate, the SN7408 AND gate, and the SN7432 OR gate all have the same pin configurations. That is, the inputs to the gates are on pins 1, 2, 4, 5, 13, 12, 10, and 9. Likewise, all of the outputs are on pins 3, 6, 11, and 8 with +5 volts (V_{cc}) wired to pin 14 and ground (GND) to pin 7. Since these devices all have the same pinouts, we should be able to design an interface and the software that can test all three of these integrated circuits.

How would these devices be tested? For the moment, let's assume that you only need to test a single 2-input SN7400 NAND gate. In order to test this gate, you would have to follow the truth table for the device, generating a combination of 1s and 0s on the inputs and monitoring the output for the proper state. With a 2-input gate, there are 2^2 or four different combinations of 1s and 0s. Likewise, if four different input states are generated, there will be four output states. If we were testing an SN7430, an 8-input NAND gate, we would have 256 different combinations of 1s and 0s for the inputs and, also, 256 output states that would have to be checked.

As you might expect, it is not very practical to test one gate in an IC, at a time. Instead, it would be much better to test all four gates in the SN7400, SN7408, and SN7432 at the same time. Thus, there are four gates, each with two inputs, so an 8-bit output port will be used to output "test patterns" to the inputs of all four gates. Likewise, we will need a 4-bit input port in order to input the output states of the four gates, so that they can be checked with the appropriate truth table.

How can we tell the microcomputer which type of gate is being tested? One way to do this is to use two thumbwheel switches, which can generate the BCD (or hex) values that are between 00 and 99. Thus, the chip to be tested would be inserted into the interface, the appropriate device number would be dialed in on the thumbwheel switches, and, then, the user would press a push button to initiate the test sequence. The interface for all of these devices is shown in Fig. 4-15, and the software for the devices is listed in Examples 4-5 through 4-8.

From Fig. 4-15, you can see that the interface consists of two input ports and two output ports. One input port is used to input the four logic levels generated by the device under test (DUT), along with the logic state of the TEST push button switch. The other input port is used to input the thumbwheel switch data, which represents the type of chip being tested. The output ports are used to output the test patterns to the device under test, and the result of the test, if it can be conducted.

The software listed in Examples 4-5 through 4-8 first waits for the TEST push button to be pressed and, then, once it is pressed, causes the microcomputer to input the information from the thumbwheel switches. The microcomputer then determines if this information represents the gates that can be tested (00, 08, or 32). While checking to see if a valid number has been input, the microprocessor alters a memory address or index so that it can address the proper table for the chip being tested. This table contains the correct information that the microprocessor should input, if the chip being tested is good. Once it accesses this table of proper responses, the microprocessor outputs one of the four "standard" test patterns, inputs the result from the IC, and compares the result to the table of correct results. If the information generated by the chip matches the information in the table, the next test pattern is output. If the two data values don't match, the IC must be bad, so the "BAD" LED in the interface is turned on. If all four values in the table match the four values obtained from the DUT, then the "GOOD" LED is turned on.

As you can probably tell from the software listings, if you try to test a device that the microcomputer hasn't been programmed to test, the "CAN'T TEST" LED will be turned on. Once an IC has been tested or the microcomputer responds with CAN'T TEST, the microcomputer will have to be reset and the program started again, in order to test additional chips. The microcomputer could be programmed to debounce the TEST push button, wait for it to be released, and then go back to ICTEST in order to test another chip. This will increase the length and complexity of the software, but it is "do-able."

MEMORY-MAPPED INPUT PORTS

As you should expect, it is possible to interface input ports to the STD bus microcomputer using memory-mapped I/O techniques. Thus, you could have a memory-mapped keyboard ADC or IC tester interfaced to your microcomputer. The only difference in the interface would be the use of the MEMRQ* signal instead of IORQ*, and the use of 16 address lines in the address decoding circuitry rather than 8 lines.

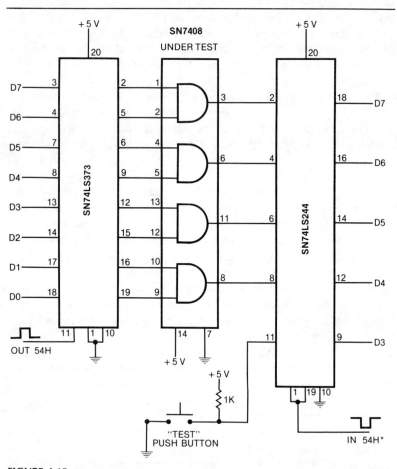

FIGURE 4-15
An interface used to test simple TTL integrated circuits.

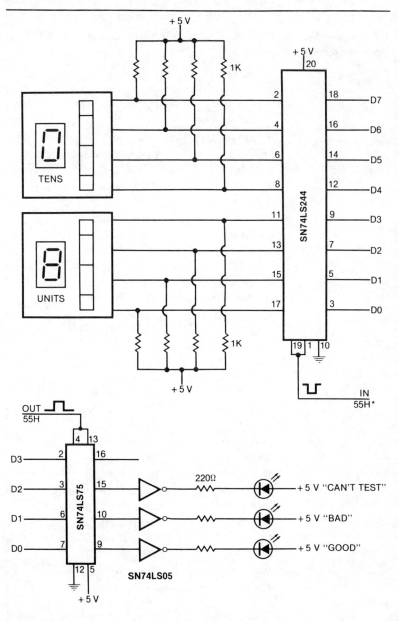

FIGURE 4-15. Cont.

Example 4-5. Simple IC Tester Software Using the 8085 Microprocessor.

ICTEST:	IN 54H	;Push button
	ANI 08H	;pressed?
	JNZ ICTEST	;No, wait for it
OK:	LXI H,ANS00	;Yes, which IC?
	LXI D,0004H	;Memory index
	IN 55H	;Get thumbwheel
	CPI 00H	;switch info
	JZ DOIT	;7400 NAND
	DAD D	;Add D&E to H&L
	CPI 08H	;7408 AND?
	JZ DOIT	;Yes
	DAD D	;No, add index
	CPI 32H	;7432 OR?
	JZ DOIT	;Yes
	MVI A,04H	;None of these,
	OUT 55H	;so turn on the
	HLT	;CAN'T TEST LED
DOIT:	LXI D,TESTP	;Load pattern address
	MVI C,04H	;# of tests
LTEST:	LDAX D	;Get a pattern
	OUT 54H	;Output to gate
	IN 54H	;Get result
	ANI F0H	;4 bits to 0
	CMP M	;Compare to
	JNZ FAIL	;proper result
	INX H	;Increment test address
	INX D	;Increment ans. address
	DCR C	;Decrement test count
	JNZ LTEST	;Loop if not 0
	MVI A,01H	;Set a bit to 1
	OUT 55H	;Turn on GOOD LED
	HLT	
FAIL:	MVI A,02H	;IC is bad
	OUT 55H	;Turn on BAD LED
	HLT	

Example 4-5 Cont.

```
TESTP:      00H         ;00000000
            55H         ;01010101
            AAH         ;10101010
            FFH         ;11111111
```

Example 4-6. Simple IC Tester Software Using the Z-80 Microprocessor.

```
ICTEST      IN A,(54H)      ;Push button
            AND 08H         ;pressed?
            JR NZ,ICTEST    ;No, wait for it
OK          LD IX,ANS00     ;Yes, which IC?
            LD DE,0004H     ;Memory index
            IN A,(55H)      ;Get thumbwheel
            CP 00H          ;switch info
            JR Z,DOIT       ;7400 NAND
            ADD IX,DE       ;Add IX to H&L
            CP 08H          ;7408 AND?
            JR Z,DOIT       ;Yes
            ADD IX,DE       ;No, add index
            CP 32H          ;7432 OR?
            JR Z,DOIT       ;Yes
            LD A,04H        ;None of these,
            OUT (55H),A     ;so turn on the
            HALT            ;CAN'T TEST LED

DOIT        LD HL,TESTP     ;Load pattern address
            LD BC,0854H     ;# of tests
LTEST       LD A,(HL)       ;Get a pattern
            OUTI            ;Output to gate
            IN A,(54H)      ;Get result
            AND F0H         ;4 bits to 0
            CP (IX)         ;Compare to
            JR NX,FAIL      ;proper result
            INC IX          ;Increment index
            DJNZ LTEST      ;Loop if not 0
            LD A,01H        ;Set a bit to 1
```

Example 4-6. Cont.

	OUT (55H),A	;Turn on GOOD LED
	HALT	
FAIL	LD A,02H	;IC is bad
	OUT (55H),A	;Turn on BAD LED
	HALT	
ANS00	DB F0H,F0H,F0H,00H	
ANS08	DB 00H,00H,00H,F0H	
ANS32	DB 00H,F0H,F0H,F0H	

Example 4-7. Simple IC Tester Software Using the 6502 Microprocessor.

ICTEST	LDA $C054	;Push button
	AND #$08	;pressed?
	BNE ICTEST	;No, wait for it
OK	LDX #$00	;Index = 0
	LDA $C055	;Get switch info
	CMP #$00	;7400 NAND?
	BEQ DOIT	;Yes
	INX	;Increment index
	INX	;by 2
	CMP #$08	;7408 AND?
	BEQ DOIT	;Yes
	INX	;Increment index
	INX	;by 2
	CMP #$32	;7432 OR?
	BEQ DOIT	;Yes
	LDA #$04	;No, turn on
	STA $C055	;CAN'T TEST LED
END	JMP END	;"Halt"
DOIT	LDA ANSPNT,X	;Get part of the
	STA $40	;address and save it
	INX	;Increment index
	LDA ANSPNT,X	;Get rest of address
	STA $41	;and save it
	LDY #$00	

Example 4-7. Cont.

TLOOP	LDA TESTP,Y	;Get a pattern
	STA $C054	;Output it
	LDA $C054	;Get the result
	AND #$F0	;4 bits to 0
	CMP ($40),Y	;Compare to answer
	BNE FAIL	;Branch if wrong
	INY	;Increment count
	CPY #$04	;4 tests yet?
	BNE TLOOP	;No, do another
	LDA #$01	;Turn on the
	STA $C055	;GOOD LED
ICOK	JMP ICOK	
FAIL	LDA #$02	;IC is bad,
	STA $C055	;so turn on the
ICBAD	JMP ICBAD	;BAD LED.
ANSPNT	.WOR ANS00	;Addresses where
	.WOR ANS08	;answers are
	.WOR ANS32	;stored

Example 4-8. Simple IC Tester Software Using the 6800 Microprocessor.

ICTEST	LDA A $C054	;Push button
	AND A #$08	;pressed?
	BNE ICTEST	;No, wait for it
OK	LDX TESTP	;Index = 0
	STX TEMP02	
	LDX ANSPNT	
	LDA A $C055	;Get switch info
	CMP A #$00	;7400 NAND?
	BEQ DOIT	;Yes
	INX	;Increment index
	INX	;by 2
	CMP A #$08	;7408 AND?
	BEQ DOIT	;Yes
	INX	;Increment index

Example 4-8. Cont.

	INX	;by 2
	CMP A #$32	;7432 OR?
	BEQ DOIT	;Yes
	LDA A #$04	;No, turn on
	STA A $C055	;CAN'T TEST LED
END	BRA END	;"Halt"
DOIT	LDA B #$04	;Set up test count
	LDX 0,X	
TLOOP	STX TEMP01	
	LDX TEMP02	
	LDA A 0,X	;Get a pattern
	STA A $C054	;Output it
	LDA A $C054	;Get the result
	AND A #$F0	;4 bits to 0
	INX	
	STX TEMP02	
	LDX TEMP01	
	CMP A 0,X	;Compare to answer
	BNE FAIL	;Branch if wrong
	INX	;Increment count
	DEC B	;4 tests yet?
	BNE TLOOP	;No, do another
	LDA A #$01	;Turn on the
	STA A $C055	;GOOD LFD
ICOK	BRA ICOK	
FAIL	LDA A #$02	;IC is bad,
	STA A $C055	;so turn on the
ICBAD	BRA ICBAD	;BAD LED.
TEMP01	EQU	;Addresses where
TEMP02	EQU	;answers are
		;stored

In the software, the I/O instructions would have to be replaced by appropriate memory-reference instructions. Of course, if a 6800, 6809, or 6502 microprocessor is being used, then the interface is already using memory-mapped I/O, except that the CPU cards have

Example 4-9. Data Acquisition Software for the 8085 Microprocessor.

ACQUIR:	LHLD COUNT	;Get the point count
	MOV A,H	;See if it is 0
	ORA L	
	RZ	;Yes, return
	LHLD STORE	;Get a memory pointer
DATA:	CALL ADC	;Get a data value
	MOV M,A	;Save the LSBs
	INX H	;Increment the address
	MOV M,B	;Save the MSBs
	INX H	;Increment the address
	PUSH H	;Save the address
	CALL DELAY	;Delay some
	LHLD COUNT	;Get the point count
	DCX H	;Decrement count
	SHLD COUNT	;Save the point count
	MOV A,H	;See if it is 0
	ORA L	
	POP H	;Get the address back
	JNZ DATA	;Not 0, get another
	RET	
DELAY:	PUSH H	;Save H & L
	LHLD TIME	;Get time delay
D1MS:	LXI D,068H	;Time constant
D11:	DCX D	;Decrement it
	MOV A,D	;Get the MSBs
	ORA E	;OR with LSBs
	JNZ D11	;Loop until 0
	DCX H	;Decrement count
	MOV A,H	;Get the MSBs
	ORA L	;OR with LSBs
	JNZ D1MS	;Delay again
	POP H	;Restore H & L
	RET	
ADC:	STA START	;Start the ADC

Example 4-9. Cont.

```
ADC1    LDA FLAG    ;Get flags and MSBs
        ANI 80H     ;Save just the flag
        JNZ ADC1    ;Wait for a 0
        LDA FLAG    ;Get MSBs again
        ANI 03H     ;Save the MSBs
        MOV B,A     ;MSBs in register B
        LDA LSB     ;Get the 8 LSBs
        RET
```

been designed to make the CPU card appear as if it is doing accumulator I/O. Thus, only the 8085, 8088, NSC800, and Z-80 have two distinct I/O modes. If one of these later-named processors is used, we would have to use different I/O instructions.

As an example, let's assume that a 10-bit memory-mapped ADC has been interfaced to the microcomputer, and we need a general-purpose data acquisition program. This program should be able to acquire any number of data points and store them in consecutive memory locations. We should also be able to adjust the time between ADC conversions, using some sort of time-delay subroutine. Since only the 8085-type microprocessors have two distinct modes of I/O operations, the software in Examples 4-9 and 4-10 is just for the 8085 and the Z-80 microprocessors.

As you can see, the software to do this is relatively simple. We have assumed that a COUNT (the number of data points) and a TIME (the number of 1 msec periods between data points) have already been loaded into the appropriate "word" memory locations. Once this is done, the ACQUIR subroutine is called. The microprocessor first checks to make sure that a count of 0 does not exist when the subroutine is called, and if the count is 0, the microprocessor immediately returns. The memory address where the data is to be stored is then loaded into registers H and L, followed by the ADC subroutine being called. The data in the A and B registers is then stored in memory and the DELAY subroutine is called. Once the microprocessor returns from the delay subroutine, the COUNT stored in memory is decremented and, if the result is not zero, another point is acquired. If the COUNT is zero, the microprocessor returns from the ACQUIR subroutine.

Example 4-10. Data Acquisition Software for the Z-80 Microprocessor.

ACQUIR	LD HL,(COUNT)	;Get the point count
	LD A,H	;See if it is 0
	OR L	
	RET Z	;Yes, return
	LD HL,(STORE)	;Get a memory pointer
DATA	CALL ADC	;Get a data value
	LD (HL),A	;Save the LSBs
	INC HL	;Increment the address
	LD (HL),B	;Save the MSBs
	INC HL	;Increment the address
	PUSH HL	;Save the address
	CALL DELAY	;Delay some
	LD HL,(COUNT)	;Get the point count
	DEC HL	;Decrement count
	LD (COUNT),HL	;Save the point count
	LD A,H	;See if it is 0
	OR L	
	POP HL	;Get the address back
	JR NZ,DATA	;Not 0, get another
	RET	
DELAY	PUSH HL	;Save H & L
	LD HL,(TIME)	;Get time delay
D1MS	LD DE,068H	;Time constant
D11	DEC DE	;Decrement it
	LD A,D	;Get the MSBs
	OR E	;OR with LSBs
	JR NZ,D11	;Loop until 0
	DEC HL	;Decrement count
	LD A,H	;Get the MSBs
	OR L	;OR with LSBs
	JR NZ,D1MS	;Delay again
	POP HL	;Restore H & L
	RET	
ADC	LD (START),A	;Start the ADC

Example 4-10. Cont.

```
ADC1    LD A,(FLAG)     ;Get flags and MSBs
        AND 80H         ;Save just the flag
        JR NZ,ADC1      ;Wait for a 0
        LD A,(FLAG)     ;Get MSBs again
        AND 03H         ;Save the MSBs
        LD B,A          ;MSBs in register B
        LD A,(LSB)      ;Get the 8 LSBs
        RET
```

The memory-mapped I/O instructions for the ADC can be seen in the ADC subroutine. The first instruction starts the ADC by writing the content of the A register out to the memory location assigned to START. It really doesn't matter what is written out to the interface because the interface does not latch this information. Instead, it simply uses the "memory-write" pulse to start the ADC. After the ADC is started, the microcomputer has to be programmed to wait for the converter's flag to go back to a logic 0, which it does in the next three instructions—starting at ADC1. Note that the "memory location" (really an input port) that contains the flag (in bit D7), also contains, in bits D0 and D1, the two most significant bits from the ADC. The eight LSBs from the converter are input by reading from the LSBs' memory location.

As you can see, the basic ADC subroutine is pretty much the same regardless of whether memory-mapped or accumulator I/O is used. Unfortunately, some of the ADC boards that are commercially available will limit you to either memory-mapped *or* accumulator I/O, so you really don't have much choice in terms of how the software has to be written.

CONCLUSION

In this chapter, you saw how input ports are designed. Input ports must use three-state or tri-state devices so that information is only placed on the microcomputer's data bus when the microcomputer requests it. The pulse that controls the three-state device is usually generated by gating together an address—either IORQ* or MEMRQ* and RD*.

Once an input port has been constructed, it can be used to input many different types of information into the microcomputer. Some of the input devices that were described in this chapter include ASCII keyboards, analog-to-digital converters (ADCs), thumbwheel switches, and simple push-button switches. As you saw, an input port can be used in an IC tester. The information that is generated by the IC under test can then be input and compared to the expected result. As was the case with output ports, the number of applications that can use input ports is unlimited.

Interrupts and Direct-Memory Access
Chapter 5

In Chapter 4, you saw how flags were used to tell the microcomputer that a peripheral needed servicing. This "service" might consist of reading a value from an analog-to-digital converter (ADC) or from a keyboard. It is also possible for output devices, such as a paper-tape punch or a floppy disk, to use flags. In both cases, a flag is used to indicate to the microcomputer that another data value has to be output so that it can either be punched on paper tape or written onto the floppy disk.

Suppose that we have a number of peripherals interfaced to the microcomputer: an ADC, a multiplexed LED display, and a high-speed line printer. If we need to use all of these devices simultaneously, for example, if an analog waveform needs to be "digitized" with the result being displayed, stored in memory in a file, and, also, printed on a line printer, the software can get very complex. The basic problem is that we would have to display information on the LED display continuously and, at the same time, monitor the flags for the ADC and the line printer. The timing for this type of software can get very complex, and it is often difficult to figure out exactly where and when to monitor the flags so that the other operations being performed by the microcomputer are not affected. Also, the microcomputer must not get "locked up" in a loop waiting for one flag to change, because the other flags may change and certain operations may not be performed, or they may be performed in the wrong order.

This problem can be solved simply by using an interrupt with the LED display, the ADC, and the high-speed line printer. What is an interrupt? An interrupt is a signal that the peripheral generates and which is wired right to the microcomputer chip (via one of the bus lines). Note that a flag signal is similar to an interrupt signal, except that the flag signal is input via the data bus and is tested with software instructions. Thus, the speed of the microprocessor determines how fast a flag can be detected.

An interrupt signal is different because it is usually buffered and is wired right to the CPU chip without using the data bus. This signal tells the CPU chip that a peripheral needs immediate servicing, just as a simple flag tells the CPU that a peripheral needs servicing. However, no software instructions have to be executed in order for the interrupt to get the attention of the CPU.

BASIC INTERRUPT OPERATION

Suppose that you are relaxing at home with the latest best seller when the doorbell rings. This interrupts your reading, so you finish the sentence that you were reading, mark your place, and answer the door. In this case, when you hear the doorbell, you are "vectored" to the door because you associate the doorbell with the front door.

In some houses, this can lead to problems, because there are a number of doors that all ring the same bell. In this case, once the "interrupt" occurs, you "poll" the doors by going from one door to the next until you finally find the person who rang the bell.

In other houses, the doorbell generates different sounds for different doors. One ring could mean the front door and two rings the back. Thus, you would be vectored to the proper door and would not have to poll them.

There is also the possibility that the doorbell rings, but you don't find anyone at the door. This happens because the person ringing the bell gets too impatient and leaves before you can answer the door. Perhaps you wanted to finish the paragraph before leaving your book.

After servicing the interrupt (you help whoever is at the door), you return to your book *and continue reading from where you left off.* At this point, the doorbell rings again, so you mark your place and answer the door. What happens if the telephone rings? You will probably tell the person at the door to wait a minute, answer the phone, and ultimately return to the person at the door. At some point, you finally get to return to your book. There are three other interrupt possibilities:

1. Ignore the interrupt—If the doorbell rings while you are reading, you may choose to ignore the interrupt (the doorbell).
2. High priority interrupts low priority—If the telephone rings (a high-priority device) while you are at the door (a low-priority device), you interrupt servicing the door to service the phone. Once the phone has been serviced, you return to the low-priority interrupt device—the doorbell.
3. Low-priority interrupt ignored—If the doorbell rings (a low-priority device) while you are on the phone (a high-priority device), the doorbell will not be answered until you *have completed* your conversation on the phone.

As you have probably already guessed, microcomputer interrupts work in much the same manner. Once an interrupt occurs, the microcomputer may have to *poll* a number of peripherals in order to determine which one caused the interrupt. Microcomputer interrupts are also *vectored*. That is, the peripherals, or internal CPU logic, tell the CPU where to find the instructions in memory which must be executed to service the device. Not only are interrupts vectored but they are also assigned a priority. If two interrupts occur at the same time (telephone and front door), some piece of hardware, or software, will determine which one of the two (or more) devices will be serviced first.

Likewise, if the microcomputer is servicing a high-priority device, such as a floppy disk, it will ignore interrupts from low-priority devices, such as a paper-tape reader or ASCII keyboard. If the microcomputer is servicing a low-priority device, it can be interrupted by a high-priority device. Once the high-priority device has been serviced, the microcomputer will return to the interrupt-service subroutine of the low-priority device and will continue to service that device.

STD BUS INTERRUPTS

Because there are a number of different CPU chips that are available on STD bus CPU cards, there is a lot of variation, in terms of interrupt capabilities, between CPU cards, even when the same CPU chip is used. However, all of the CPU cards have the NMIRQ* interrupt input. This is the highest priority-interrupt input to the CPU card, and it is a *nonmaskable interrupt*. That is, it can never be masked out or disabled. The other interrupt input, INTRQ*, can be disabled by executing one assembly language instruction.

The NMIRQ* interrupt is a vectored interrupt; the vector is generated by internal CPU hardware. The memory locations that the CPUs are vectored to are summarized in Table 5-1. As you can see, the 8085-type CPUs are vectored to low memory and the 6800-type CPUs are vectored to high memory. From Fig. 5-1, you can see that it is very easy to interrupt an STD bus microcomputer with the NMIRQ* line.

Table 5-1. Nonmaskable Interrupt (NMIRQ*) Vector-Memory Locations

Processor	Memory Location	Content
8085	0024H	an instruction
Z-80	0066H	an instruction
NSC800	0066H	an instruction
8088	00008H*	an address
6800	FFFCH	an address
6809	FFFCH	an address
6502	FFFAH	an address

*The 8088 processor can address 1,048,576 bytes of memory; thus, the use of a 5-digit hex address.

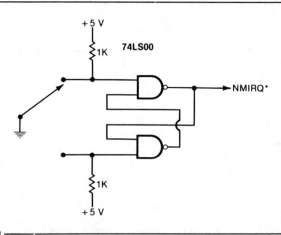

FIGURE 5-1

Circuit for using the nonmaskable interrupt input (NMIRQ*) to interrupt the microcomputer.

In Fig. 5-1, a dpst switch and some NAND gates are all that is required to use the NMIRQ* interrupt input. The NAND gates simply remove glitches from the switch closure. Once the microcomputer is interrupted, it is vectored to the section of memory that contains either (1) the interrupt service instructions (8085, Z-80, and NSC800) or (2) the address where the instructions are stored (8088, 6800, 6809, and 6502). Since the nonmaskable interrupt is often used for catastrophic problems (power failure), the interrupt service instructions are used to save the content of registers and temporary values and, then, shut the microcomputer down in an orderly manner. The microcomputer would then halt. Once power was restored, the microcomputer would be reset and would continue from where it left off.

There is only one other interrupt input to the CPU card, INTRQ*. One of the differences between NMIRQ* and INTRQ* is that the INTRQ* input can be disabled through software. If this input is disabled, the microcomputer will not be interrupted by a logic 0 on this line of the bus. Usually, this input to the CPU card is not *latched*. This means that if the interrupt is disabled and a logic 0 pulse occurs on the INTRQ* line (from a logic 1 to a logic 0, and back to a 1), the microcomputer will not remember that a peripheral needed servicing, once the interrupt is enabled.

One signal that the CPU card generates, once it has been interrupted, is interrupt acknowledge, or INTAK*. Usually, this signal is not generated when an interrupt occurs on the NMIRQ* line. Of course, there is no way of knowing exactly when the microprocessor is going to be interrupted by INTRQ*. The microprocessor could be right in the middle of executing an instruction. Thus, the interrupt will *only* be acknowledged after the microprocessor has finished executing the current instruction.

At the same time that the interrupt is acknowledged, the interrupt is automatically disabled. Thus, no other devices can interrupt the microprocessor, with the exception of the NMIRQ* device. If a low-priority device is being serviced, higher-priority devices should be able to interrupt the servicing of this device. Thus, in low-priority interrupt service subroutines, the interrupt would be re-enabled by one of the very first instructions in the subroutine. If a high-priority device is being serviced, the interrupt would only be enabled after the device is serviced. This prevents low-priority devices from interrupting the servicing of the high-priority device.

Let's return to the ADC/LED/high-speed printer application and see how these signals are used with these peripherals. As you know, each one of these devices has to generate an interrupt. Of course, both the ADC and printer have flags to tell the processor that they either have or need data. In the case of the LED display, which doesn't have a flag, a 2-msec one-shot multivibrator (monostable; SN74121 or SN74123) would have to be added to the interface. This device would be triggered whenever information is output to the display, and the flag would go to a logic 0 two milliseconds later. One of the flags could be wired to the NMIRQ* and one to the INTRQ* interrupt input, but there is no third interrupt input for the third peripheral. In practice, each one of these devices would be wired to the INTRQ* line, and the NMIRQ* line would be reserved for future use. In order to wire a flag from each device to the INTRQ* line, *open-collector ICs* would have to be used. These ICs are readily available and are inexpensive. The problem is, once an interrupt occurs, *how can we determine which device caused the interrupt if they are all wired to the same interrupt input?* Because each of the microprocessors available for use with the STD bus varies greatly in terms of how interrupts are handled, each microprocessor will be discussed either individually or by functional group.

THE 8085

When the 8085 microprocessor was designed, Intel Corporation decided that *each interrupting peripheral would supply* a *unique vector to the microprocessor.* This vector is actually a *single-byte call instruction,* and the vector is gated into the instruction register of the 8085. When this instruction is executed, a return address is saved on the stack. This return address is actually the address of the *next instruction* in the main program that *would have been* executed had the interrupt not occurred. Thus, when the NMIRQ* interrupt is used, an internal vector is used to alter the program counter, and when the INTRQ* is used, the peripheral must provide a vector. There are eight of these vectors or *restart instructions.* They are summarized in Table 5-2.

How is this 8-bit vector actually gated into the microprocessor chip? When the interrupt is acknowledged with INTAK*, the three-state logic on the peripheral board is enabled (just like an input port), and the information is placed on the data bus. The microprocessor knows that this information, since it was placed on the data bus dur-

Table 5-2. 8085 Vectors (Restart Instructions) and Their Memory Locations

Name	Value Hex	Value Binary	Memory Location
RST0	C7	11000111	0000H (Same as RESET)
RST1	CF	11001111	0008H
RST2	D7	11010111	0010H
RST3	DF	11011111	0018H
RST4	E7	11100111	0020H
RST5	EF	11101111	0028H
RST6	F7	11110111	0030H
RST7	FF	11111111	0038H

ing an interrupt-acknowledge cycle, should go into the instruction register rather than a general-purpose register. This vector, or instruction, is then executed. The 8085 microprocessor "calls" one of eight possible interrupt-service subroutines and, in the process, saves a return address on the stack. Once the peripheral has been serviced, a return instruction at the end of the interrupt-service subroutine causes the return address to be popped off of the stack and into the program counter. The net effect is that the 8085 processor continues executing the program that was interrupted—from the point where the interrupt occurred.

From Table 5-2, you can see that the addresses of the interrupt-service subroutines that can be called are only 8 memory locations apart. Since many interrupt-service subroutines are 20 or more instructions long, we cannot store much of a subroutine in only 8 memory locations. The solution to this problem is to store a jump instruction in each one of these "vector" memory locations. By using a jump instruction, control will be transferred to another section of memory, where as much or as little memory as is necessary can be used for each interrupt-service subroutine.

One problem with this technique is that the peripheral-interface cards start to get very complex. Not only does the interface have to transfer data between the peripheral and the CPU, but it also has to have an 8-bit interrupt port which will place an 8-bit vector on the data bus during the interrupt-acknowledge cycle. To solve this problem and to keep the peripheral-interface cards as simple as possible, a single parallel interrupt-controller card (PICC) is used in each system.

In an 8085 system, this would be the only card that was wired to INTRQ* and INTAK*. The flags from each of the peripherals that needed to generate an interrupt would be wired to a user connector on the PICC, opposite the 56-pin edge connector. The INTAK* signal would also be used by this card, and it would cause one of eight possible vectors to be placed on the data bus.

What will happen if two devices try to interrupt the 8085 at exactly the same time? Of course, the PICC can't place two vectors on the data bus at the same time, because the 8085 wouldn't be able to figure out what to do. The solution to this problem is actually solved by hardware on the PICC. Each peripheral has to *be assigned a priority by the user,* and the priority is *established by the position of the flag signal on the user edge connector on the PICC.* Thus, if two devices try to interrupt the 8085 at the same time, *the vector of the higher priority device will be placed on the data bus.*

How does the PICC do this? It really doesn't matter if we understand exactly how it is done, but it is possible to do this with an SN74LS148 priority encoder chip (Fig. 5-2). From the truth table (Fig. 5-2B), you can see that Input 7 has the highest priority and Input 0 has the lowest priority. If Input 7 is a logic 0, regardless of the states of the other inputs, the outputs of the chip (A2, A1, and A0) will be in the logic 0 state. The outputs are only in this state when Input 7 is a logic 0. Thus, not only does the priority encoder chip determine the priorities, but it also generates a unique value (vector) for each and every input.

If you reexamine Table 5-2, you can see that all of the 8-bit vectors are very similar. In fact, bits D7, D6, D2, D1, and D0 are always at logic 1. The three other bits, D5, D4, and D3, change to reflect the eight different possible vectors. By using the SN74LS148 with an 8-bit three-state buffer, a very simple PICC can be designed (Fig. 5-3). Additional circuitry is really needed to get this to work properly, but at least you get the idea. Remember, you won't be designing PICCs; you'll buy a card from one of the manufacturers.

As you can see, the PICC establishes a priority between all eight possible interrupt devices. The reason that this is called a parallel interrupt controller is because all eight peripheral flags are wired, in parallel, to the interrupt controller. It would also be proper to call the interrupt controller a parallel priority interrupt controller, although all of the interrupt controllers that we know of establish a priority. When the Z-80 microprocessor is described, a serial technique will be discussed.

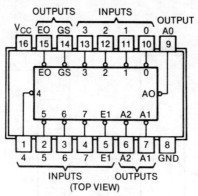

(A) Pin configuration.

			INPUTS							OUTPUTS			
E1	0	1	2	3	4	5	6	7	A2	A1	A0	GS	EO
H	X	X	X	X	X	X	X	X	H	H	H	H	H
L	H	H	H	H	H	H	H	H	H	H	H	H	L
L	X	X	X	X	X	X	X	L	L	L	L	L	H
L	X	X	X	X	X	X	L	H	L	L	H	L	H
L	X	X	X	X	X	L	H	H	L	H	L	L	H
L	X	X	X	X	L	H	H	H	L	H	H	L	H
L	X	X	X	L	H	H	H	H	H	L	L	L	H
L	X	X	L	H	H	H	H	H	H	L	H	L	H
L	X	L	H	H	H	H	H	H	H	H	L	L	H
L	L	H	H	H	H	H	H	H	H	H	H	L	H

(B) Function table.

FIGURE 5-2

The SN74148/SN74LS148 priority encoders. *(Courtesy Texas Instruments Incorporated)*

THE 8088

The 8088 (also designed by Intel Corporation) is very similar to the 8085. If an 8088 is used, there are 256 possible vectors. The values of the vectors range from 00H to FFH. Like the PICC developed for the 8085, these 8-bit values are gated onto the data bus during INTAK*. *However, these values are not gated into the instruction register; they are not signal-byte call instructions.* Instead, the vector is multiplied by four and the result is used to address memory. The resulting address will be in the range of 00000 through 003FC. From this section of memory, the 8088 processor will read two 16-bit values. One value is loaded into the program counter and the other value is loaded into the code segment, or CS register. The net result is that

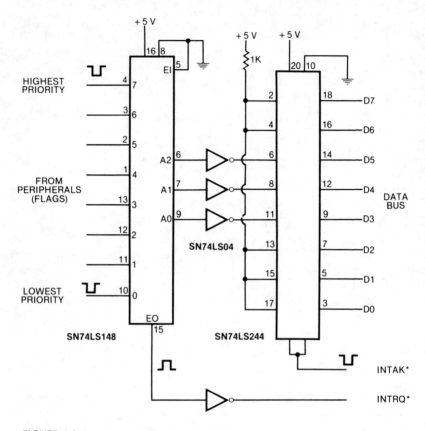

FIGURE 5-3
A simple parallel priority interrupt controller.

these two values are used to address the next instruction that is to be executed. In the 8085 processor, the vector was an instruction that was executed. In the 8088, the vector *points to a section of memory that contains the addresses of all of the interrupt-service subroutines.*

As always, a return address is saved on the stack (both the PC and CS registers), so that when the 8088 returns from the interrupt-service subroutine by executing the IRET instruction, these values are popped off of the stack. They are loaded back into the proper registers and the 8088 processor continues from where it left off. Having the interrupt vector point to, or address, a table of interrupt-service subroutine addresses is very common, and is one of the techniques available on the Z-80 and NSC800 microprocessors.

Even though the PICC that was designed previously generated single-byte vector instructions for the 8085 microprocessor, this exact same card can be used on 8088-based STD bus systems. The same vectors will be generated, except that the 8088 will treat them as pointers to a table of addresses. Table 5-3 summarizes the memory addresses that are addressed by these same vectors.

Table 5-3. 8088 Memory Locations Addressed by 8085 Vectors

8085 Vector	Memory Location Addressed
C7H	0031CH
CFH	0033CH
D7H	0035CH
DFH	0037CH
E7H	0039CH
EFH	003BCH
F7H	003DCH
FFH	003FCH

THE Z-80 AND NSC800

The Z-80 and NSC800 microprocessors have identical instruction sets and respond to the NMIRQ* and INTRQ* signals in the same manner. The only real difference between these two CPU chips is that they have different pinouts and the NSC800 draws less current than the Z-80 (10 mA vs. 150 mA). Both of these microprocessors have very powerful interrupt handling capabilities.

If you are using the Z-80 or the NSC800 processor, you may have skipped the last two sections on the 8085 and 8088. If you did, go back and read them because the same information is applicable to both the Z-80 and the NSC800. From this point on, we will just use the term Z-80, although we really mean both processors.

When the Z-80 microprocessor was designed, it was decided that it should be able to execute the same instructions as the 8080 processor—the predecessor of both the Z-80 and 8085. It was also designed to handle interrupts in the same manner. At the same time, additional instructions were added to the Z-80's instruction set, and two additional modes of interrupt operation were added. Thus, all 8080 programs can be executed on a Z-80-based microcomputer. The three modes of interrupt operation are called Mode 0, Mode 1, and Mode 2. The user sets the mode of interrupt operation by executing one of three possible two-byte assembly language instructions.

In Mode 0, the Z-80 processor acts just like an 8085. This means that one of the vectors in Table 5-2 can be placed on the data bus during INTAK*. The Z-80 will execute this single-byte instruction and call a subroutine in low memory. The "standard" PICC can be used with the Z-80 microprocessor when it is operating in this mode.

In Mode 1, any pulse on the INTRQ* line causes the Z-80 microprocessor to be vectored to memory location 0038H. Thus, the INTRQ* line can be used just like NMIRQ*, except that we can enable and disable the INTRQ* input. This mode is not frequently used.

The most powerful mode is Mode 2. It is very similar to the 8088's interrupt structure. In this mode, the user can place any one of 128 possible 8-bit vectors on the data bus, during INTAK*. This vector is used in conjunction with the internal I register of the Z-80 microprocessor to address a block of 256 memory locations. These memory locations contain 128 16-bit memory addresses for 128 interrupt-service subroutines. This is shown in Fig. 5-4.

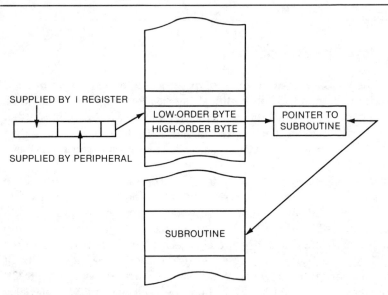

FIGURE 5-4

Using the I register in the Z-80 or NSC800 microprocessors (along with a 7-bit peripheral vector) to generate a 16-bit memory address. *(Courtesy National Semiconductor Corp.)*

Since the I register in the CPU chip is 8-bits wide, it provides the 8 most-significant address bits for the 16-bit table address. Thus, the

256 memory locations used to store the table must all have the same upper 8 bits of address. The table could be stored in memory from 3000H through 30FFH but not from 4C54H through 4D44H. Since only the low 8 bits of the address can change, there can be a maximum of 128 addresses stored in these 256 memory locations. Thus, there are only 128 vectors that can be used. This also means that bit A0 of the 16-bit memory address must be 0, as seen in Fig. 5-4.

When a peripheral actually takes INTRQ* to a logic 0, the interrupt is acknowledged and the INTAK* signal is used to gate an 8-bit vector on the data bus, where D0 is 0. This is combined with the content of the I register and the resulting memory address is used to read from memory, into the program counter, the address of the interrupt-service subroutine for the peripheral that generated the vector. Once the peripheral has been serviced, the Z-80 microprocessor returns from the interrupt-service subroutine and continues to perform the task that was interrupted.

SERIAL PRIORITY

When the 8085 microprocessor was discussed, a parallel interrupt controller was used, where flags from all of the peripherals were wired to the edge connector on the controller card. With the Z-80 and NSC800 microprocessors, a serial priority technique can be used, where the peripheral card closest to the CPU has the highest priority and the priority of all other peripheral cards decreases the further away from the CPU card that they are. Thus, the position of the card in the system determines its priority.

This serial priority technique can only be used in the Z-80 and NSC800 systems, because the MOS LSI peripheral chips that have been designed for use with the Z-80 processor have this serial interrupt logic designed into them. It is too cumbersome to design this hardware with standard TTL parts, so no one does. If the serial priority technique is used in your Z-80-based STD bus microcomputer, then two additional signals on the STD bus are used—PCO and PCI.

The priority chain output (PCO) and priority chain input (PCI) signals are unlike any other STD bus signals. That is, all of the PCO pins (pin 51) are not connected together and all of the PCI pins (pin 52) are not connected together. If you look at the physical arrangement of these two signals (Table 1-1), you will see that they are on opposite sides of the edge connector. Thus, in a rack system, the PCO signal of one edge connector is connected to the PCI pin of the adjacent connector. The PCO signal on this adjacent connector is connected to the PCI pin on the next edge connector, and so on. If you want to verify this, just look at pins 51 and 52 on your system's motherboard.

As you might guess, the priority chain output (PCO) signal is generated by peripheral cards and the priority chain input (PCI) signal is used by peripheral cards. The four possible logic states of these two signals are summarized in Table 5-4. From this table, you can see that if a peripheral interrupts the microcomputer, the PCI input to the card will be a logic 1 and the PCO output will be a logic 0. The INTAK* signal would then be used by the Z-80 MOS LSI interface chip located on this board to place the 8-bit vector that is stored in the chip on the data bus. Of course, the user has to store the appropriate vector in the interface chip before interrupts are used.

Table 5-4. Logic States of the PCI and PCO Interrupt Signals

PCI	PCO	Indication
0	0	Higher priority interrupt being requested.
0	1	Undefined and not allowed.
1	0	Interrupt device has highest priority and is being serviced.
1	1	No interrupt.

How does all of this establish a priority between peripheral devices? Let's use a high-speed printer (highest priority), an ADC (medium priority), and an LED display (lowest priority) as an example. If both the high-speed printer and the LED display try interrupting the Z-80 microcomputer at the same time, the PCO and PCI signals from and to the LED display interface will both be logic 0. The PCI line to the high-speed printer interface will be a logic 1 and the PCO line from the printer interface will be a logic 0. This logic 0 prevents lower-priority devices from interrupting the servicing of the high-speed printer (the highest priority device). This combination of PCI and PCO (1 and 0) also determines which interface will use INTAK* to place a vector on the data bus.

Another possibility is that the microcomputer is servicing a low-priority device when a high-priority interrupt occurs. The result is shown in Fig. 5-5. Basically, the microcomputer will be interrupted by the high-priority device, right in the middle of servicing the low-priority device. Once the high-priority device has been serviced, the

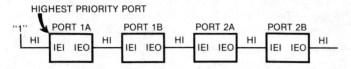

(A) Priority interrupt daisy chain before any interrupt occurs.

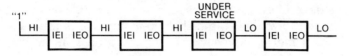

(B) Port 2A requests an interrupt and is acknowledged.

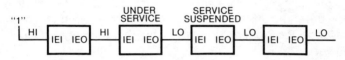

(C) Port 1B interrupts and suspends servicing of Port 2A.

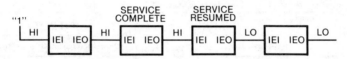

(D) Port 1B service routine complete, "RETI" issued, and Port 2A is resumed.

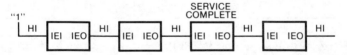

(E) Second "RETI" instruction is issued on completion of Port 2A service routine.

FIGURE 5-5

A high priority device interrupting the servicing of a low-priority device. *(Courtesy Mostek Corp.)*

microcomputer will return to servicing the low-priority device. Once the low-priority device has been serviced, the microcomputer will return to the original interrupted task.

THE 6800, 6809, AND 6502

The 6800-type microprocessors all handle interrupts in much the same manner. At the beginning of this chapter, you saw that when a nonmaskable interrupt request occurs, internal hardware generates a vector to high memory. From this section of memory, all three microprocessors read the 16-bit address of the nonmaskable interrupt instructions. The same is true for the INTRQ* input.

When the INTRQ* goes to a logic 0, internal hardware causes the 16-bit address in the program counter to be pushed onto the stack. In addition, the 6502 processor causes the flags to be saved on the stack. On the 6800 and 6809 microprocessors, all of the other registers (accumulators, index registers, etc.) including the flags are saved on the stack. Once these values have been saved, the microprocessor reads a memory address in high memory into the program counter. This points the microprocessor to the instructions in the interrupt-service subroutine that have to be executed.

The memory locations that are used by both the nonmaskable and maskable interrupts are summarized in Table 5-5. As you can see, since both the 6800 and 6809 were developed by Motorola Semiconductor Products, Inc., the same four memory locations have been used.

Table 5-5. Interrupt Memory Locations Used by the 6800, 6809, and 6502 Microprocessors

	NMIRQ*		INTRQ*	
	High Byte	Low Byte	High Byte	Low Byte
6800	FFFC	FFFD	FFF8	FFF9
6809	FFFC	FFFD	FFF8	FFF9
6502	FFFA	FFFB	FFFF	FFFF

Since there is only one INTRQ* input, the microprocessor has to determine which one of the many possible interrupt-driven devices caused the interrupt. If the 8085-type processor is used, the device provides a vector on the data bus during interrupt acknowledge. However, the 6800 and 6502 processors do not generate an interrupt-acknowledge signal (INTAK*) and they do not expect a vector on the data bus. The 6809 does generate the INTAK* signal, but it doesn't expect a vector on the data bus either.

In order to find out which device generated the interrupt, the 6800-type microprocessors have to poll a flag port. The flag port is simply an 8-bit input port, with the inputs to this port wired to the flags of the interrupting devices. These same flags are wired to the INTRQ* line in the bus by using "open-collector" chips. In the interrupt-service subroutine, the microprocessor polls these flags by reading the input port into a register and examining the 8-bit value on a bit-by-bit basis to find the device that requested service. The priority of the various devices is determined by (1) the position of the device's flag in the 8-bit input port value, and (2) the order used to examine the bits in the 8-bit value. A simple hardware design that could be used to input the flag values is shown in Fig. 5-6. Like the PICC design shown previously, additional hardware is required to make this circuit completely functional.

In Fig. 5-6, you can see that the flags from the peripheral interface cards are wired to both INTRQ* and the interrupt-controller card. This interrupt-controller card is very simple; it is just an 8-bit input port with the inputs to the input port wired to the user edge connector. Once the 6800-type microprocessor is executing the instructions in the interrupt-service subroutine, this 8-bit value can be input into a register where the bits can be tested. In order to test these bits for the appropriate logic 1 or logic 0 state, the value could be shifted or rotated into the Carry or Sign bit, and then the bit would be "tested" with a conditional branch instruction.

6809 IMPROVEMENTS

When Motorola Semiconductor Products, Inc. designed the 6809 microprocessor, one improvement that they added to the CPU chip was the ability to generate an interrupt-acknowledge signal. This signal is generated by gating the logic 0 state of the bus acknowledge (BA) signal with the logic 1 state of the bus status (BS) signal. *The resulting signal can be used to gate a peripheral-specific memory address onto the data bus (as 2 bytes), when the 6809 reads from "memory locations" FFF8 and FFF9.* To do this requires a fairly sophisticated PICC, along with a well-designed 6809 CPU card.

If interrupts are polled, the microprocessor wastes valuable time. It is much better (and more efficient) if peripherals provide vectors to the microprocessor, as is done with the 8085, Z-80, NSC800, and 8088. However, whenever the INTRQ* line goes to a logic 0, the 6809 processor will always read a 16-bit memory address from memory locations FFF8 and FFF9. In most systems, this section of memory is EPROM, because when the 6809 is reset, it reads the address of the first instruction to be executed, from memory locations FFFE and FFFF. If this was a Read/Write memory, the microprocessor would read a random reset address from these two memory locations.

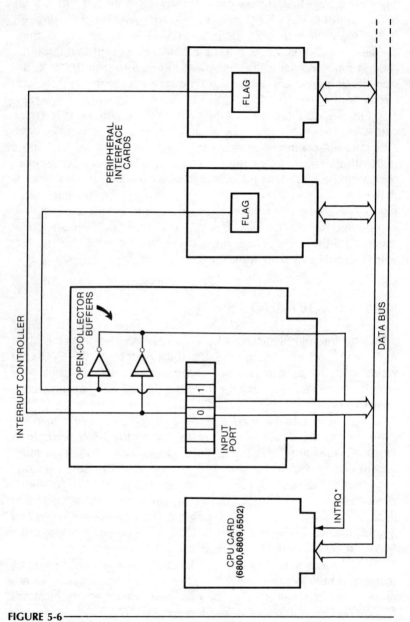

FIGURE 5-6

Using an input port to read flags from interrupt-driven devices.

In order for each peripheral or interface to provide the 6809 processor with a peripheral-specific vector, *this section of memory must be disabled by the INTAK* signal.* At the same time, a peripheral-specific address can be placed on the data bus, a byte at a time, by the peripheral. One way of doing this is shown in Fig. 5-7. Like the other interrupt controllers discussed, this figure does not contain all of the required circuitry.

The PICC now contains 16 bytes of R/W memory so that the 16-bit vectors (memory addresses) for 8 devices can be stored. This memory is selected, and the memory on the CPU card is deselected, when the interrupt-acknowledge signal (INTAK*) is a logic 0. The address for this R/W memory is generated by a combination of A0 from the CPU card (it will be 0 when a Read from FFF8 is performed, and a 1 for FFF9) and the output of the SN74LS148 priority encoder. If device 7 (the highest priority device) generated the interrupt, the outputs of the SN74LS148 will be logic 0s, so memory locations 0 and 1 contain the vector for device 7, the highest priority device.

Before any interrupts occur in the system, the user would first have to save 16 bytes of information in this R/W memory. Once this is done, the interrupt can be enabled. Note that the present schematic does not show any way in which information can be written out into this memory. However, it is not important that we worry about this and other details. Remember, you will probably buy a PICC, not build one.

AN INTERRUPT REVIEW

It may be a little difficult for you to digest all of this new information, so let's take it from the top. As soon as the microcomputer is reset, the user has to ensure that the microcomputer will respond to interrupts correctly. In some microcomputers, the PICC may have to be initialized, some values may have to be output to the Z-80 peripheral chips, some tables of addresses may have to be moved around in memory, etc. Once this is done, the interrupt can be enabled. At some later time, a peripheral generates an interrupt by taking the INTRQ* line to a logic 0. The serial- and parallel-priority hardware ensures that the highest priority device is always serviced first.

Once the microcomputer detects the interrupt and acknowledges it, if it can, the interrupt is disabled. Various internal registers will be

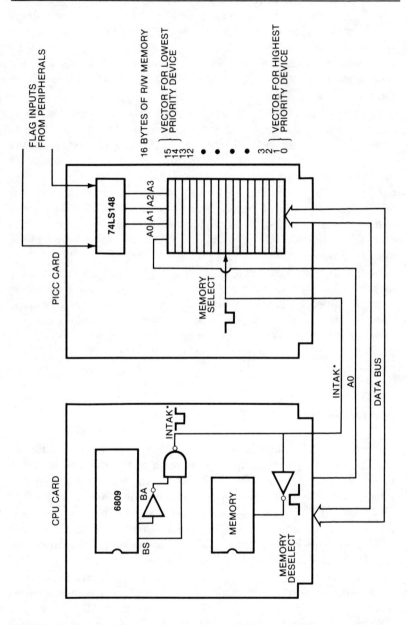

FIGURE 5-7
Generating peripheral-specific vectors for a 6809 STD bus system.

saved on the stack along with a return address—the address of the next instruction that would have been executed, if the interrupt hadn't occurred.

By using the INTAK* signal, many of the microprocessors can place a peripheral-specific vector on the data bus. This external logic, or some internal logic in the case of the 6800 and 6502 processors, alters the content of the program counter so that the microprocessor starts executing the appropriate interrupt-service subroutine. At some point in the interrupt-service subroutine, the interrupt must be reenabled. For high-priority devices, the interrupt would be reenabled just before the microprocessor returns from the interrupt-service subroutine. This prevents low-priority devices from interrupting the servicing of the high-priority device. If a low-priority device is being serviced, then the interrupt would be reenabled at the beginning of the interrupt-service subroutine. By doing this, a high-priority device can interrupt the servicing of the low-priority device—which is the way the system should work.

Once the peripheral has been serviced, the microprocessor needs to return to the interrupted task. On some microprocessors, the microprocessor simply has to execute a subroutine-return instruction. On other microprocessors, a special return-from-interrupt instruction has to be executed. These special return instructions are used either to pop a number of registers off of the stack or to initialize external interrupt hardware.

INTERRUPT SOFTWARE

In order to show you what interrupt software looks like, let's use an analog-to-digital converter (ADC), a multiplexed LED display, and some type of counter/timer chip. The ADC is an 8-bit 25 μsec analog-to-digital converter, where a Write operation to port 00H starts the converter (C000H for memory-mapped I/O) and the 8-bit conversion can be read from port 00H (C000H). The flag of the ADC is wired to the INTRQ* line via a PICC or serial-interrupt hardware.

It really doesn't matter what type of counter/timer chip is used, as long as it can be interfaced to the microcomputer using interrupts. We have also assumed that the counter/timer chip can be programmed to generate a series of pulses. Like the ADC flag, the counter/timer flag is wired to INTRQ* with some additional hardware.

What do we want the microcomputer to do? Using the ADC and

the counter/timer, an analog signal must be digitized at the rate of 1000 samples/sec, for 5 seconds. Thus, the microcomputer needs to be programmed to acquire and store 5000 8-bit values in memory. In addition, while all of this is going on, the microcomputer will be displaying information on a 10-digit multiplexed LED display.

The software that does all of this is combined in Examples 5-1 through 5-4. Regardless of the type of microcomputer being used, a memory pointer and count have to be established first, followed by the programming and starting of the counter/timer and the enabling of the interrupt. Again, it really doesn't matter what it takes to get the counter/timer going, as long as it will generate an interrupt every 1 msec. Once this is done, the microcomputer can display information on the LED display. It doesn't matter what type of information is being displayed as all we wanted to do was to keep the microprocessor busy doing something. This is the task that will be interrupted every 1 msec. Note that there are no flag-monitoring instructions in any of the software examples.

When an interrupt occurs, the 8085 and Z-80 will immediately begin servicing the proper device because vectored interrupts have been used. In the 6502 and 6800 examples, the microprocessor has to poll the interrupt flag port to determine which device (the counter/timer or the ADC) caused the interrupt.

The first interrupt that occurs will be due to the timer since the ADC has not been started. In the timer interrupt-service subroutine, some registers are saved simply because we can't predict exactly what the microprocessor was doing when the interrupt occurred. Thus, we don't know which registers were being used just before the interrupt occurred. After the registers are saved (the 6800 processor saves them automatically), the count has to be decremented.

If the result of this operation is nonzero, the count has to be saved (one way or another), the ADC has to be started, the registers have to be restored to their preinterrupt value, and the interrupt has to be reenabled. Remember, when an interrupt occurs, the interrupt is automatically disabled. Once the ADC has been started, an interrupt will occur in 25 μsec—the conversion time of the ADC.

If the count has been decremented to zero, all of the data points have been acquired, so there is no point in starting the ADC. Instead, the timer may have to be stopped, depending on the microprocessor chip in the system, and, then, the registers have to be restored to their preinterrupt state. In the 6800 and 6502 examples, the timer has to be stopped because the interrupt is reenabled by the RETI instruction.

Example 5-1. Servicing a Counter/Timer and an ADC Using an 8085 Microprocessor.

```
            ORG 2000H
START:      LXI SP, STACK     ;Load the SP
            LXI H,BUFFER      ;Set up an address
            SHLD PNTR         ;Save it
            LXI H,COUNT       ;Set up a count
            SHLD CNT          ;Save it
              .               ;Program and start
              .               ;timer (1 ms)
            EI                ;Enable interrupt
MP10:         .               ;Drive the display
              .
              .
            JMP MP10

            ORG 0008H
            JMP TIMER         ;RST1 vector

            ORG 0010H
            JMP ADC           ;RST2 vector

TIMER:      PUSH PSW          ;Save A and flags
            PUSH H            ;Save H and L
            LHLD CNT          ;Get point count
            DCX H             ;Decrement it
            SHLD CNT          ;Save point count
            MOV A,H           ;Get MSBs of count
            ORA L             ;OR with LSBs
            JNZ DOADC         ;Not 0, do an ADC
                              ;Got points, stop timer
            POP H             ;Restore H and L
            POP PSW           ;Restore A and flags
            RET

DOADC:      OUT 00H           ;Start the ADC
EXIT:       POP H             ;Restore H and L
            POP PSW           ;Restore A and flags
            EI                ;Enable interrupt
            RET
```

Example 5-1. Cont.

ADC:	PUSH PSW	;Save A and flags
	PUSH H	;Save H and L
	LHLD PNTR	;Get memory address
	IN 00H	;Get ADC value
	MOV M,A	;Save in memory
	INX H	;Increment pointer
	SHLD PNTR	;Save new pointer
	JMP EXIT	;Restore registers

Once the ADC has completed the conversion, another interrupt will occur. This time, the microprocessor will execute the instructions in the ADC interrupt-service subroutine. At this point, registers are saved, the value is read from the ADC and stored in memory, the memory pointer is incremented (and saved), and the interrupt is re-enabled. The next interrupt that occurs will be the timer interrupt, which occurs about 970 μsec after the ADC has been serviced.

INTERRUPTS AND THE STACK

As you already know, a return address is saved on the stack when the microcomputer is interrupted. Some additional information is also saved on the stack automatically, depending on the CPU chip that

Table 5-6. Information Automatically Saved on the Stack When an Interrupt Occurs

CPU Chip	Information
8085	PC
Z-80	PC
NSC800	PC
8088	PC, CS, flags
6800	PC, A, B, X, flags
6502	PC, flags
6809	PC, U, X, Y, DP, A, B, flags

PC = program counter; CS = code segment; X,Y = index registers; A,B = accumulators; U = user stack; DP = direct page

Example 5-2. Servicing a Counter/Timer and an ADC Using the Z-80 Microprocessor.

```
              ORG 2000H
START         LD SP,STACK        ;Load the SP
              LD A,30H
              LD I,A
              IM 2               ;Interrupt mode 2
              LD HL,BUFFER       ;Set up an address
              LD DE,CNT          ;Set up a count
              EXX                ;Exchange register bank
                .                ;Program and start
                .                ;timer (1 ms)
              EI                 ;Enable interrupt
MP10:           .                ;Drive the display
                .
                .
              JP MP10

              ORG 3000H
TABLE         DEFW TIMER         ;RST1 vector
              DEFW ADC           ;RST2 vector

TIMER         EXX                ;Exchange registers
              EXX AF,AF'         ;Save A and flags
              DEC DE             ;Decrement point count
              LD A,D
              OR E
              JR NZ,DOADC        ;Not 0, do an ADC
                                 ;Stop timer
              EXX AF,AF'         ;Exchange A and flags
              EXX                ;Restore registers
              RETI               ;Return from interrupt

DOADC         OUT (00H),A        ;Start the ADC
EXIT          EXX AF,AF'         ;Exchange A and flags
              EXX                ;Exchange registers
              EI                 ;Enable interrupt
              RETI
```

Example 5-2. Cont.

ADC	EXX AF,AF[1]	;Save A and flags
	EXX	;Exchange registers
	IN A,(00H)	;Get memory address
	LD (HL),A	;Get ADC value
	INC HL	;Save in memory
	JP EXIT	;Increment pointer
		;Save new pointer
		;Restore registers

you are using. The information that is saved on the stack by the various processors is summarized in Table 5-6.

On the 8085, 8088, and 6502 microprocessors, the general-purpose registers and the index registers should be saved on the stack at the beginning of the interrupt-service subroutine. Since an interrupt can occur at any time, there is no way of knowing which registers were in use when the interrupt occurred. On the Z-80 and NSC800 processors, the register exchange instructions (EX AF,AF' and EXX) can be executed so that the interrupt-service subroutine uses the alternate set of registers. If these registers are used in the main program, the registers would have to be pushed on the stack. Since the 6800 and 6809 microprocessors automatically save all of the registers on the stack, the user doesn't have to execute any "register save" instructions in the interrupt-service subroutine.

Once the instructions in the interrupt-service subroutine have been executed, it is important that the user return all of the registers to their preinterrupt state. If the user pushed registers onto the stack, then all of the same registers will have to be popped off of the stack—in the proper order. On the Z-80 and NSC800 processors, the register exchange instructions would be executed. On the 6800, 6809, and 6502 processors, there are special return-from-interrupt instructions (RTI) that pop all of the information off of the stack. The Z-80 also has a special return-from-interrupt instruction (which is used to reset the interrupting peripheral), as does the 8088 (so that both the PC and CS are popped off of the stack). If you are using the 8085, you can use an ordinary subroutine-return instruction (RET).

Example 5-3. Servicing a Counter/Timer and an ADC Using a 6502 Microprocessor.

START	LDS #$FF	;Load the SP
	LDA #$89	;Part of 500
	STA $40	;(Number of
	LDA #$13	;points)
	STA $41	
	LDA #$00	;Save address
	STA $42	;on page 0
	LDA #$40	;also
	STA $43	
	.	;Program and start
	.	;the timer (1 ms)
	CLI	;Enable interrupts
MP10	.	;Drive the display
	.	
	.	
	JMP MP10	
	ORG $FFFE	
	DW SERVIC	
SERVIC	PHA	;Save accumulator
	LDA FLAGS	;Get flag byte
	BMI ADC	;Is it the ADC?
TIMER	DEC $40	;No, must be the timer
	BNE DOADC	;Decrement count
	DEC $41	;If not 0, start the
	BNE DOADC	;ADC, otherwise,
		;stop the timer
	PLA	;Restore accumulator
	RTI	
DOADC	STA $C000	;Start the ADC
	PLA	;Restore accumulator
	RTI	
ADC	STY $44	;Save Y
	LDY #$00	;Load Y with 0
	LDA $C000	;Get ADC value

Example 5-3. Cont.

```
              STA ($42),Y      ;Save it in memory
              INC $42          ;Increment the
              BNE SKIP         ;memory address
              INC $43
    SKIP      LDY $44          ;Restore Y
              PLA              ;Restore A
              RTI
```

INTERRUPT TIMING

The last interrupt topic that needs to be discussed is *interrupt timing*. This means that you need to be careful when deciding how frequently the microcomputer is to be interrupted, and how long it takes for the microcomputer to execute the interrupt-service subroutines.

As an example, let's use the previous data-acquisition (ADC/timer) example. Of course, if you are using a 25-μsec ADC instead of a 10-msec ADC, you won't be able to acquire a data point every 1 msec. This is obvious, but it is easy enough to forget some of the small details when you start using a number of interrupt devices in a system.

Suppose that the timer and the ADC are the two lowest priority devices in the system. There are also a few other peripherals that interrupt the microcomputer on an irregular basis. The ADC has a conversion time of 25 μsec (as before) and the timer generates an interrupt every 1 msec (as before). Problems will occur if a higher priority interrupt occurs and it takes more than 1 msec to service this higher priority device.

If the microcomputer takes 5 msec to service this high-priority device, 5 data points will be missed, simply because the 1-msec interrupt from the timer was ignored since it was a lower priority device. Unfortunately, you may not be able to determine that data points were missed simply by looking at your data. However, in some cases, it will be obvious that data points were missed.

As you might expect, even if the highest priority device only needs 200–300 μsec to be serviced, data points may still be missed because other high-priority devices may also interrupt the microcomputer so that the total time to service all of the higher prior-

Example 5-4. Servicing a Counter/Timer and an ADC Using a 6800 Microprocessor.

```
START       LDS STACK           ;Load the SP
            LDX POINTER         ;Set up data POINTER
            STX PNTR
            LDX COUNT           ;Set up point counter
            STX CNTR
                .               ;Program
                .               ;timer
            CLI                 ;Enable interrupts
MP10            .               ;Drive the display
                .
                .
            JMP MP10

            ORG $FFFE
            DW SERVIC


SERVIC      LDA A FLAGS         ;Get flag byte
            BMI ADC             ;Is it the ADC?
TIMER       LDX CNTR            ;No, must be the timer
            DEX                 ;Decrement count
            STX CNTR            ;If not 0, start the
            BNE DOADC           ;ADC, otherwise,
                                ;stop the timer
            RTI

DOADC       STA A $C000         ;Start the ADC
            RTI

ADC         LDA A $C000         ;Get ADC value
            LDX PNTR            ;Get memory pointer
            STA A 0,X           ;Save it in memory
            INX                 ;Increment the address
            STX PNTR            ;Save it

            RTI
```

ity interrupts will be more than 1 msec. On the other hand, if acquiring data from the ADC on a regular basis is the most important task that your microcomputer has to perform, then the ADC and the timer should be the highest priority devices in the system. This may mean that you will have to eliminate some peripherals and change the priority of others.

You also have to ensure that no device can interrupt the servicing of itself. This means that if the timer generates an interrupt every 1 msec, it must not take 2–3 msec to service the timer. Otherwise, once the first timer interrupt occurs, the timer servicing will be interrupted by a timer interrupt and the return addresses and registers will start to accumulate on the stack. The end result is that the microcomputer will never return from the interrupt-service subroutine and, eventually, your program will "bomb out."

Even though some commercially available PICCs can handle between eight and sixteen interrupt-driven devices, you should be very cautious about having this many interrupt-driven devices, simply because the software and timing gets very complex, very quickly. The more interrupt-driven devices that you have in a system, the easier it is for an interrupt to interrupt the servicing of an interrupt. As before, the microcomputer can become *interrupt bound,* so that it never has a chance to finish the servicing of one device before another interrupt occurs.

DIRECT-MEMORY ACCESS

There are a number of cases where even using interrupts does not get data into the microcomputer, or out of the microcomputer, as fast as the peripheral needs to operate. There are many peripherals, including disks, fast ADCs (500-nsec conversion time), and some physical measurement transducers that can either produce data faster than the microcomputer can input and store it in memory, or will need information faster than the microcomputer can output it. Assume that information needs only to be input from the peripheral. At a minimum, the microcomputer has to input the data, store it in memory, increment a memory address (or index), decrement a count, and loop back if the count is nonzero. All of this takes too much time for some fast peripherals, so data will be lost.

The solution to this may be to use the direct-memory access (DMA) technique, where information is passed between memory and the peripheral directly without having to go through one of the microprocessor's internal general-purpose registers for intermediate storage. In order to be able to perform DMA, the peripherals' interface has to mimic the operation of the microprocessor chip on the CPU card. This interface has to generate a 16-bit memory address which is placed on the address bus, data has to be placed on the data bus, and the MEMRQ* along with either, or both, RD* and WR* have to be controlled.

REQUESTING THE BUS

Ordinarily, the CPU card is the card that controls these signals. At the same time, the DMA card can't simply place an address, data, and control signals on the buses, because the DMA controller and the CPU would both be trying to use the same buses at the same time. The solution to this is to use the bus request (BUSRQ*) and bus acknowledge (BUSAK*) signals. To use these signals, the DMA controller would take BUSRQ* to a logic 0. This tells the CPU card that it must stop using the buses, so that the DMA controller can begin using them.

Once the buses are requested, the CPU will force the circuitry on the CPU card that drives the buses into the high-impedance state, and will then generate the BUSAK* signal. At this point, the CPU card is no longer placing information on these buses. (In fact, to the buses, the CPU card looks like it doesn't even exist.) Therefore, the DMA controller can enable its three-state address, data, and control bus drivers. After information is transferred between the DMA peripheral and memory, the DMA controller sets its bus drivers into the high-impedance state and takes the BUSRQ* signal back to a logic 1. This causes the CPU circuitry to be enabled so that an address, data, and the control signals are placed on the buses by the CPU card and processing can continue.

On most systems, there is a delay of a few μsec between the time that the buses are requested (BUSRQ*) and the BUSAK* signal is generated. If any interrupt (NMIRQ* or INTRQ*) occurs during a DMA transfer, it is ignored. If an interrupt is being serviced, a DMA operation can occur right in the middle of the interrupt-service subroutine. Thus, DMA operations have a higher priority than interrupts. *Extreme care must be used when doing DMA transfers in systems that use dynamic read/write memories.* Most STD bus systems that use dynamic memories are based on the Z-80 or the NSC800 microprocessor. During a DMA transfer, these two CPU chips cannot generate the refresh information required by this type of memory. If long DMA transfers occur (more than 2 msec), information will be lost from this type of memory. However, if static read/write memory is being used, there won't be any trouble.

DIRECT-MEMORY ACCESS CONTROLLERS

In most cases, you won't be building a DMA controller. Many of the semiconductor manufacturers (Intel, AMD, Mostek, Zilog, NEC, and Motorola) have designed MOS LSI DMA controllers for use with their CPU chips. Thus, you will probably be able to buy a card that contains one of these chips if you need to do DMA. A typical DMA controller chip can control up to four DMA devices and can transfer up to 65,536 bytes of information between the DMA device and memory. Quite often, when you purchase a high-speed peripheral, such as a floppy disk or a fixed-head disk, the manufacturer will be able to supply a direct-memory access controller for your STD bus system.

As we mentioned, the MOS LSI DMA controller chips can often handle up to four DMA devices and, like interrupts, there has to be a priority between the DMA devices. There are usually two different priority modes built into these chips: fixed priority and rotating priority. The fixed priority mode of operation uses the fixed priority that was established when the chip was designed. In some chips, channel No. 4 has a higher priority than channel No. 3, etc., and this can't be changed. In the rotating priority mode of operation, once the highest priority DMA device has been serviced, it becomes the lowest priority device and the priority of all the other channels are "bumped up" one.

At the beginning of this chapter, a serial-interrupt technique was discussed that is used with the Z-80 microprocessor. A serial priority DMA daisy chain is also possible. However, at this moment, it has not been designed into any STD bus cards because there simply aren't any traces in the bus that could be dedicated to this function. All 56 conductors have been used.

Realistically, you will not have 10–12 DMA devices in your system. In most cases (99% or more), one DMA controller chip that can handle four DMA devices is all that is needed. Again, if you should need a lot of DMA, maybe you should try using two or three smaller systems.

DMA SOFTWARE

Since DMA transfers are all performed without any intervention by the CPU, there is no DMA software like there is with interrupt and I/O software. Of course, if an MOS LSI DMA controller is used, then you will have to have assembly language instructions in your program that program the controller with the memory address(es) to be used, the number of bytes to be transferred, decision as to whether a Read or Write operation is to take place, and whether fixed or rotating priority is to be used. Once one of these DMA chips has performed the DMA transfer, it will generate an interrupt and/or set a bit in an input port to indicate to the CPU that the DMA transfer has been completed. Software may also be used to initiate the transfer or tell the peripheral when to begin the transfer.

Some very simple software that can be used with a simple DMA device has been listed in Example 5-5. In these programs, the microcomputer simply reads the content of a memory location and outputs this value to an LED output port. Ordinarily, we would not expect to see the value that is being displayed change. However, by using a DMA controller, you could write directly into the memory location being "monitored" and you would see the lights change. Each time the DMA controller requested the buses, wrote a value into memory, and then caused the buses to be relinquished, the displayed value would change. Note that the displayed value only changes once the buses have been relinquished so that the program can read the value from memory and display it. If the buses are never relinquished (some type of hardware problem), then the program won't be executed and the value won't change.

CONCLUSION

In this chapter, we have discussed two very different techniques for transferring information between the microcomputer and a peripheral. In Chapter 4, you saw how flags were used to notify the microcomputer that a peripheral needed servicing. The only problem with this technique is that it is slow and the microcomputer has to periodically monitor these flags. However, if interrupts are used, the microcomputer doesn't have to monitor any flags. Instead, when the peripheral needs data or has data, it interrupts the microcomputer and the data are then transferred when the interrupt-service subroutine is executed.

Example 5-5. Monitoring and Displaying a DMA Memory Location.

1. Using an 8085 microprocessor:

LOOP	LDA DMALOC	;Get memory value
	OUT 34H	;Output to LEDs
	JMP LOOP	;Do it again

2. Using a Z-80 microprocessor:

LOOP	LD A,(DMALOC)	;Get memory value
	OUT (34H),A	;Output to LEDs
	JR LOOP	;Do it again

3. Using a 6502 microprocessor:

LOOP	LDA DMALOC	;Get memory value
	STA $C034	;Output to LEDs
	JMP LOOP	;Do it again

4. Using a 6800 microprocessor:

LOOP	LDA A DMALOC	;Get memory value
	STA A $C034	;Output to LEDs
	BRA LOOP	;Do it again

There are a number of new questions that now have to be asked when using interrupts (including serial or parallel priority interrupts), when using vectors or polling the peripheral's interrupt flags, when using interrupt-service subroutines or saving registers or deciding when to reenable the interrupt, or when deciding on the speed at which interrupts can be serviced, and when deciding on the need to program the interrupt controllers. Thus, new decisions have to be made also.

When very high-speed peripherals are interfaced to the microcomputer, direct-memory access (DMA) techniques may have to be used. This technique doesn't use any software to transfer data; it is all done in hardware. It is used by the highest speed peripherals and the transfer rate is only limited by the MOS LSI DMA controller used and the access time of the memories in the system. Transfer rates of 1–2 Mbytes/sec are common.

General-Interest Interface Cards
Chapter 6

In this chapter, some of the commercially available interface cards will be discussed. In the following sections, the function of a card will be described, along with the jumpers that must be used to make the card work. Some software examples will be given. Thus, this chapter answers the question, "What do I have to do to make the card work in my system?"

One of the basic selection criterion used (when selecting the cards to be discussed) was that the cards had to be STD bus compatible, and not just STD-Z80, STD-6800, etc., compatible. Thus, all of the cards described in this chapter can be used with any of the available CPU cards and systems.

THE MOSTEK DIOB/DIOP

In many systems, the STD bus microcomputer is used to control high-power/high-voltage devices, such as 110 V ac lights, motors, and relays. On the other hand, the microcomputer is often used to sense switch closures, where there is 24 or 48 V dc across the switch. These voltages are common when controlling mixers, pumps, electric valves, hoists, and motors.

To control these devices requires some special modules, such as the ones called solid-state relays and solid-state input modules. The solid-state relays are designed to control ac and dc devices, at up to 220 V and 10–20 A, using a single TTL input. The input modules convert an ac or dc signal to a TTL signal. Since both types of devices are TTL compatible, it is easy to interface these devices to a microcomputer, using output ports to drive the solid-state relays and using input ports with the solid-state input modules. Because these two functions (input and output) are required in many systems, there are a number of different cards that are available for your use.

There are also a number of different ways in which the solid-state relays and input modules can be wired to the output ports and the inputs ports. The simplest design contains both the solid-state control devices and I/O ports on the same standard-size STD bus card. A user edge connector is provided so that the motors (110 V ac, etc.) can be connected directly to the card. Typical cards contain 8 solid-state relays or 8 input modules.

This configuration has one basic limitation. The more solid-state control modules that you need, the more cards you will need to plug into your system. If you need 30 solid-state relays and 10 input modules, you will need 6 cards which will use up 6 slots in your system. This may cause you to have to upgrade your system and get a larger motherboard.

There is a variation of this design, however, where each interface board still controls 8 solid-state modules, but the modules are located in a separate enclosure or rack, and the TTL control lines flow between the interface card and the rack of relays and input modules through a flexible cable.

The Mostek system is exceptional, in that (1) you can have any mix of input modules and solid-state relays, and (2) you can control up to 256 modules using a *single* interface board. This system is shown in Fig. 6-1. It consists of one digital I/O bus (DIOB) interface board and up to sixteen of the digital I/O panels (DIOP). The boards are interconnected by a 40-conductor flexible cable.

The digital I/O bus (DIOB) card is basically a bus extender. That is, the 8-bit data bus and address lines A0–A5, along with MEMRQ* gated with both RD* and WR*, are present on the user edge connector. These signals are wired through the 40-conductor cable to each one of the DIOPs. In addition, circuitry on the DIOB compares A6–A15 of a memory address to an address that is jumpered on the board by the user. Only when the user-jumpered address matches the memory address on the address bus will information flow through the interconnection cable. Since this board has been designed to use a 16-bit address, it can only be used as a *memory-mapped I/O device.* Also, since A0–A5 are wired to all of the digital I/O panels, this card uses up 64 consecutive memory addresses even though only one rack of 16 solid-state modules is used.

(A) Digital I/O bus interface board.

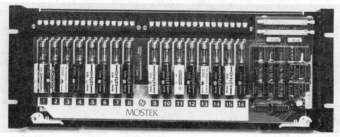

(B) Digital I/O panel.

FIGURE 6-1

The Mostek DIOB/DIOP industrial control system. *(Courtesy Mostek Corp.)*

The jumpers that are used to determine which block of memory addresses the DIOB responds to are shown in Fig. 6-2. In order to configure the DIOB for addresses C800 through C83F, which jumpers would be needed? It will be necessary to have jumpers 5–6, 7–8 (for the 0s in the "C"), 11–12, 13–14, and 15–16 (for the three 0s in the "8"), and 17–18 and 19–20 for the two 0s in the hex values between 0 and 3. Jumpers are added to the board for the logic 0s in an address (lack of a jumper will give you a logic 1).

The DIOP board is just as easy to use. Since A0–A5 are wired to each of the DIOPs, there are additional decoders and comparators on each board. Each board can be used with any combination of sixteen input and output modules, so this requires two port addresses. In addition, two more port addresses are used for data direction registers (output ports). The pattern of 1s and 0s in the data direction registers

197

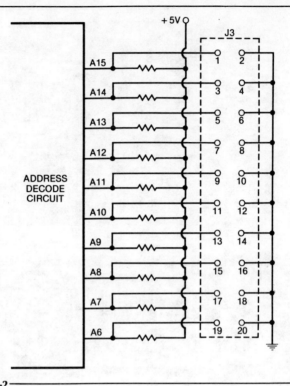

FIGURE 6-2

Jumper A6 through A15 for the DIOB interface card. *(Courtesy Mostek Corp.)*

determines whether a bit in one of the solid-state module ports will be used to control an input module or an output module. Thus, each DIOP uses up four port addresses. Since A0–A5 are wired to all DIOP cards, there are 2^6, or 64, possible port addresses. Each DIOP uses four addresses, so there can be a maximum of sixteen DIOP cards in a system.

There are only four jumpers that the user has to add to the DIOP. These are shown in Fig. 6-3. These jumpers are used just like the jumpers on the DIOB. If a jumper is added to the board, a logic 0 results; if no jumper is used, a logic 1 results. The sixteen possible combinations for these jumpers is summarized in Table 6-1, along with the DIOP card that is selected. As you can see from Fig. 6-3, jumpers are not added to the DIOP for either A5 or A0.

In order to select one of the two 8-bit ports (data or direction) on the DIOP, bit A0 is used. Bit A5 is used to determine if one of the I/O

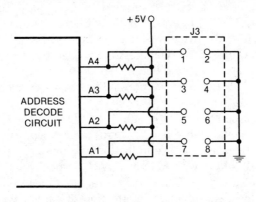

FIGURE 6-3
The A1–A4 jumpers that are located on the digital I/O panel. *(Courtesy Mostek Corp.)*

Table 6-1. Jumpers to Select One of 16 DIOP Cards

Pins 1–2	Pins 3–4	Pins 5–6	Pins 7–8	DIOP Selection Number
0	0	0	0	1
0	0	0	1	2
0	0	1	0	3
0	0	1	1	4
0	1	0	0	5
0	1	0	1	6
0	1	1	0	7
0	1	1	1	8
1	0	0	0	9
1	0	0	1	10
1	0	1	0	11
1	0	1	1	12
1	1	0	0	13
1	1	0	1	14
1	1	1	0	15
1	1	1	1	16

0 = CLOSED, 1 = OPEN   *Courtesy Mostek Corp.*

ports is being accessed (A5 = 0) or if one of the data direction ports is being accessed (A5 = 1). Thus, the first DIOP in the system might use addresses C800 and C801 for the I/O module ports, and C820 and C821 for the data direction ports. What jumpers would have to be added to the DIOP to make it respond to these addresses? All four jumpers (1–2, 3–4, 5–6, and 7–8) would have to be added, so that address bits A1–A4 in the 16-bit memory address are all 0s. Remember, part of the memory address is decoded by the DIOB (block of 64 memory addresses) and part by the DIOP (four of 64 addresses). Also, each DIOP uses four memory addresses—two where A5 is a logic 0, and two where A5 is a logic 1.

The different combinations of A0, A5, CRD, and CWR are shown in Table 6-2, along with the ports that are selected. The CRD and CWR control signals are active in the logic 1 state and are generated by NANDing MEMRQ* with RD* and WR*, respectively. The two illegal Read states for the data direction registers mean that once we have output the data direction information, it cannot be read back into the microcomputer. This is normally not a problem. Also, note that the CRD and CWR signals are active in the logic 1 state. If they were active in the logic 0 state, they would be called CRD* and CWR*. As you face the DIOP, with the flat cable connectors at the upper right-hand side of the board, the left port is on your left (module positions DK1–8) and the right port is on your right (module positions DK9–16).

Table 6-2. Reading From and Writing to the DIOP

A5	A4	A3	A2	A1	A0	CRD	CWR	Operation
0	X	X	X	X	0	1	0	Read left data port
0	X	X	X	X	0	0	1	Write left data port
0	X	X	X	X	1	1	0	Read right data port
0	X	X	X	X	1	0	1	Write right data port
1	X	X	X	X	0	1	0	NOT ALLOWED
1	X	X	X	X	0	0	1	Write left data direction port
1	X	X	X	X	1	1	0	NOT ALLOWED
1	X	X	X	X	1	0	1	Write right data direction port

CRD and CWR are active in the logic 1 state.

In their Operations Manual, Mostek Corp. uses the DIOB and DIOP to control a conveyor belt. The conveyor belt is driven by a motor that is started with a START push-button switch, and can be stopped by using any one of 7 STOP switches distributed along the conveyor belt. There is also a RUNNING status light and a STOPPED status light. In order to control this device, three output modules are needed (one for the motor and two for the lights), and eight input modules (seven stop switches and one start switch) are needed.

Since only eleven devices need to be controlled, a single DIOP can be used, with the left port being used for the output devices and the right port being used for the input devices. The assignment of the switches, lights, and motor to the various bits present in each port is given in Table 6-3.

Table 6-3. Assignment of Lights, Switches, and Motor in the DIOP

Port	Channel	Data Bit	Function
Left	DK1	D7	RUNNING light; starts motor
	DK2	D6	STOPPED light
	DK3	D5	Unused
	DK4	D4	Unused
	DK5	D3	Unused
	DK6	D2	Unused
	DK7	D1	Unused
	DK8	D0	Unused
Right	DK9	D7	START push button, normally open, momentary
	DK10	D6	HALT switch, normally closed
	DK11	D5	HALT switch, normally closed
	DK12	D4	HALT switch, normally closed
	DK13	D3	HALT switch, normally closed
	DK14	D2	HALT switch, normally closed
	DK15	D1	HALT switch, normally closed
	DK16	D0	HALT switch, normally closed

Left = C0C0; Right = C0C1

The basic operations that the microcomputer has to perform in order to control the conveyor belt can be seen in the flowchart of Fig. 6-4. First, the microcomputer checks to see if the conveyor belt is running. If it is running, then the microcomputer has to see if any of the HALT switches are closed. If none of these switches are closed, then the microcomputer has nothing else to do. If one of the HALT switches is closed, the RUNNING light is turned off (which means the conveyor belt motor is turned off) and the STOPPED light is turned on. At this point, the microcomputer is done.

If the conveyor belt is not running, the microcomputer also checks the state of the seven HALT switches. If one of them is closed, the microcomputer exits. This means that if one of the HALT switches is

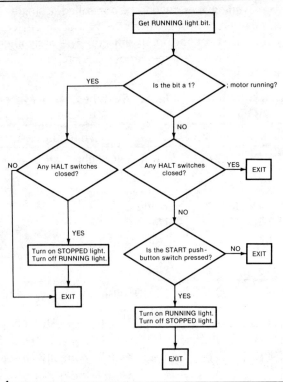

FIGURE 6-4
Flowchart of conveyor belt operation.

in the HALT position, the motor cannot be started. The switch has to be placed back in the RUN position, and then the motor can be restarted. If none of the HALT switches are closed, the microcomputer checks the state of the START push-button switch. If it is pressed, the RUNNING light is turned on (which also causes the motor to be started) and the STOPPED light is turned off.

The software that can be used to control the conveyor belt is listed in Examples 6-1 through 6-4. The INIT instructions are used to initialize the data direction registers in the DIOP. Remember, a 1 is used for an output bit and a 0 is used for an input bit. The "zero" bits in the left port are unused, and they are programmed as input bits. Once the DIOP is programmed, the microcomputer can perform other tasks.

Periodically, the CHECK software is executed. In this software, the microprocessor *reads* the state of the RUNNING output bit to determine if the motor is running or not. If it is running, the microprocessor jumps to RUNNG. If the motor is not running, the states of the HALT switches are checked. If one of them is a logic 0, the microprocessor exits because the motor isn't running; one of the HALT switches is closed. If none of the HALT switches are closed and the

---

Example 6-1. Controlling a Conveyor Belt with a Mostek DIOB/DIOP Using an 8085 Microprocessor.

INIT:	LXI H,00C0H	;11000000 = left
	SHLD 0C0E0H	;00000000 = right
	.	
	.	
	.	
CHECK:	LDA 0C0C1H	;Get switch information
	MOV B,A	;Save it in B
	LDA 0C0C0H	;Get light status
	ORA A	;Set the CPU flags
	JM RUNNG	;Motor is running
	MOV A,B	;Motor stopped, get
	ANI 07FH	;switch information
	CPI 07FH	;Any HALT switches on?
	JNZ EXIT	;Yes, so exit
	MOV A,B	;No, start motor?
	ORA A	;Set CPU flags
	JP EXIT	;Don't start, exit
	MVI A,080H	;Start the motor
	STA 0C0C0H	
	JMP EXIT	;and exit
RUNNG:	MOV A,B	;Get switch information
	ANI 07FH	;Save just the HALTs
	CPI 07FH	;Any HALTs on?
	JZ EXIT	;No, so don't halt
	MVI A,040H	;Yes, stop the motor
	STA 0C0C0H	
EXIT:		
	.	
	.	
	.	

Example 6-2. Controlling a Conveyor Belt with a Mostek DIOB/DIOP Using a Z-80 Microprocessor.

INIT	LD HL,00C0H	;11000000 = left
	LD (C0E0H),HL	;00000000 = right
	.	
	.	
	.	
CHECK	LD A,(C0C1H)	;Get switch information
	LD B,A	;Save it in B
	LD A,(C0C0H)	;Get light status
	OR A	;Set the CPU flags
	JP M,RUNNG	;Motor is running
	LD A,B	;Motor stopped, get
	AND A,07FH	;switch information
	CP 07FH	;Any HALT switches on?
	JP NZ,EXIT	;Yes, so exit
	LD A,B	;No, start motor?
	OR A	;Set CPU flags
	JP P,EXIT	;Don't start, exit
	LD A,080H	;Start the motor
	LD (C0C0H),A	
	JP EXIT	;and exit
RUNNG	LD A,B	;Get switch information
	AND A,07FH	;Save just the HALTs
	CP 07FH	;Any HALTs on?
	JP Z,EXIT	;No, so don't halt
	LD A,040H	;Yes, stop the motor
	LD (C0C0H),A	
	.	
	.	
	.	

motor isn't running, the state of the START switch is checked. If this push-button switch is being pressed, the motor is started by changing the states of the RUNNING and STOPPED lights.

If the motor is running, the microprocessor jumps to RUNNG, where the state of the HALT switches is checked. If one or more of them is closed, the motor is stopped by turning the RUNNING light off and turning the STOPPED light on.

Example 6-3. Controlling a Conveyor Belt with a Mostek DIOB/DIOP Using a 6502 Microprocessor.

INIT	LDA #$C0	;11000000 = left
	STA $C0E0	
	LDA #$00	;00000000 = right
	STA $C0E1	
CHECK	BIT $C0C0	;Motor running?
	BMI RUNNG	;Yes, so branch
	LDA $C0C1	;Get HALT switches
	AND #$7F	;Save just the HALTs
	CMP #$7F	;Any HALTs on?
	BNE EXIT	;Yes, so exit
	BIT $C0C1	;No HALTs, start motor?
	BPL EXIT	;No, so exit
	LDA #$80	;Yes, start motor
	STA $C080	
	JMP EXIT	
RUNNG	LDA $C0C1	;HALT switches
	AND #$7F	;Save HALT switches
	CMP #$7F	;Any closed?
	BEQ EXIT	;No, so exit
	LDA #$40	;Yes, so stop motor
	STA $C0C0	
EXIT	.	
	.	
	.	

THE ENLODE 214 DISPLAY SYSTEM

The Enlode 214 display system can be used to display up to 40 alphanumeric characters, using ten 4-character LED display modules (Fig. 6-5). This display system consists of two cards, one of which is plugged into the STD bus system. The other card contains the LED displays. The two cards are connected by a 44-conductor flat cable that can be up to 4 feet long.

In order to display information on the display, the user has to output a character position (00H–27H), followed by the ASCII value of

Example 6-4. Controlling a Conveyor Belt with a Mostek DIOB/DIOP Using a 6800 Microprocessor.

INIT	LDX #$C000	;11000000 = left
	STX $C0E0	;00000000 = right
	.	
	.	
	.	
CHECK	TST $C0C0	;Motor running?
	BMI RUNNG	;Yes, so branch
	LDA A $C0C1	;Get HALT switches
	AND A #$7F	;Save just the HALTs
	CMP A #$7F	;Any HALTs on?
	BNE EXIT	;Yes, so exit
	TST $C0C1	;No HALTs, start motor?
	BPL EXIT	;No, so exit
	LDA A#$80	;Yes, start motor
	STA A $C080	
	JMP EXIT	
RUNNG	LDA A $C0C1	;HALT switches
	AND A #$7F	;Save HALT switches
	CMP A #$7F	;Any closed?
	BEQ EXIT	;No, so exit
	LDA A #$40	;Yes, so stop motor
	STA A $C0C0	
	.	
	.	
	.	

the character to be displayed. Character position 00H is the character closest to the flat-cable connector, and position 27H is at the other end of the board. Normally, we would read the displayed information from character position 00H over to 27H (left to right).

Before information can be output to this display system, I/O addresses have to be assigned to the card. When the card is shipped from the factory, the character-position port address is set to E0H and the data (or ASCII) character-port address is set to E1H. If you need to use other addresses, jumpers for these addresses would have to be removed from the interface board and new ones would have to be added. This board can only be used as an accumulator I/O device.

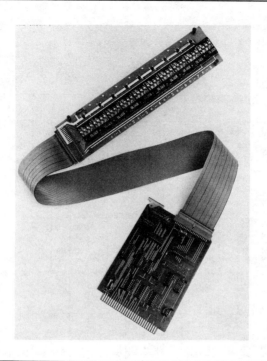

FIGURE 6-5

The Enlode 214 40-character display system. *(Courtesy Enlode Inc.)*

Thus, only four or five jumpers are needed to configure the interface card.

The addresses that are available and the jumpers that are required are summarized in Table 6-4. As you can see, there are four jumpers that are used on the board to select the two port addresses. The X and Y jumpers are wired to decoders that decode A7–A2 and the ZA and ZB jumpers are used to select which one of the four possible states of A0 and A1 will be used for the character (data) port address (ZB) and the character-position address (ZA). The physical location of the jumpers on the card is shown in Fig. 6-6.

If the card has already been jumpered for E0H (character position, ZA) and E1H (data, ZB), which jumpers have been used? From Table 6-4, jumpers would be made between X and 7, Y and 0, ZA and 0, and ZB and 1. What jumpers would be needed to use address 5A for the data port and 5B for the character-position port? The jumpers needed would be X to 2, Y to 6, ZA to 3, and ZB to 2.

Table 6-4. Jumper/Address Assignments for the Enlode Display Card

Most Significant Hex Address	Least Significant Hex Address																Jumper Section X,Y,ZA,ZB
	0	1	2	3	4	5	6	7	8	9	A	B	C	D	E	F	ZA ↓ ZB
	Z0	Z1	Z2	Z3	Z0	Z1	Z2	Z3	Z0	Z1	Z2	Z3	Z0	Z1	Z2	Z3	
0		X0 Y0		X0 Y1		X0 Y2		X0 Y3	X AND Y								
1		X0 Y4		X0 Y5		X0 Y6		X0 Y7									
2		X1 Y0		X1 Y1		X1 Y2		X1 Y3									
3		X1 Y4		X1 Y5		X1 Y6		X1 Y7									
4		X2 Y0		X2 Y1		X2 Y2		X2 Y3									
5		X2 Y4		X2 Y5		X2 Y6		X2 Y7									
6		X3 Y0		X3 Y1		X3 Y2		X3 Y3									
7		X3 Y4		X3 Y5		X3 Y6		X3 Y7									
8		X4 Y0		X4 Y1		X4 Y2		X4 Y3									
9		X4 Y4		X4 Y5		X4 Y6		X4 Y7									
A		X5 Y0		X5 Y1		X5 Y2		X5 Y3									
B		X5 Y4		X5 Y5		X5 Y6		X5 Y7									
C		X6 Y0		X6 Y1		X6 Y2		X6 Y3									
D		X6 Y4		X6 Y5		X6 Y6		X6 Y7									
E		X7 Y0		X7 Y1		X7 Y2		X7 Y3									
F		X7 Y4		X7 Y5		X7 Y6		X7 Y7									

Courtesy Enlode Inc.

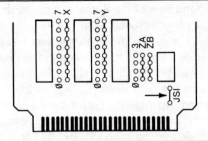

FIGURE 6-6
Jumper locations on the display interface card. *(Courtesy Enlode Inc.)*

When selecting addresses for this interface, you have to keep the address within a multiple-of-4 address boundary. Thus, you cannot jumper the board for addresses F7 and F8, 23 and 24, or 2F and 30. The least-significant digit of the address for both ports must be in the range of 0–3, 4–7, 8–B or C–F. Within this range, the character-position and data ports can be jumpered to any of these four possible addresses.

It is possible for some microprocessors to output information to the display so fast that it doesn't have time to propagate down the cable and into the displays before the next data value is output. If your microprocessor can output a character position, data, and a new character position in 3 μsec or less, you need to add jumper JS1 to the interface board. This jumper causes WAIT states to be inserted into the microprocessor's timing, when an ASCII value is written out to the display. This jumper is needed when the interface is used with a 4-MHz Z-80 microprocessor. The interface board is normally shipped without this jumper.

Because of the display modules used, it is possible to display a "cursor" at any of the 40 possible display positions. The cursor consists of turning on all 16 segments in the addressed-character position. Because of the display module's internal design, if a cursor is displayed, the character that was being displayed is saved inside the LED display module. Once the cursor is moved, the original character is automatically displayed again. Thus, you don't have to remember the character that was displayed, display a cursor, and then output the original character. All that you have to do is output the cursor and, then, erase the cursor.

To display a cursor, an 81, 82, 84, or 88 has to be output. An 81 will cause the cursor to be displayed on the right-hand display within the 4-character module and an 88 will cause the cursor to be displayed on the left-hand side of the 4-character module. A value of 8F will cause cursors to be displayed in all 4-character positions while an 80 will erase all cursors within a 4-character module.

Note that the cursor values actually contain the character position where the cursor is to be displayed. Thus, the two least-significant bits in the character-position value (00H–27H) are ignored when a cursor value is output. To place a cursor on character 3 (the right-hand character in the left-hand display module; near the cable), you could output the character position 00, 01, 02, or 03, and follow it with the cursor value 81.

What do the software drivers for this peripheral look like? The display initialization software is contained in Examples 6-5 and 6-6. The software that can be used to display a 40-character message is contained in Examples 6-7 and 6-8.

In the initialization software (Examples 6-5 and 6-6), the microcomputer has to turn off all cursors and save a blank or space (all segments off) in each of the 40 character positions. To do this, some registers are saved on the stack and, then, a register is loaded with the first character position to be accessed, 00H. In the INIT loop, the character position is output, all four cursors in the accessed display module are turned off, and a space is output to the addressed character position. Once this is done, the character position is incremented and compared to the value 28H, which is one more than the address of the last character position. Once this value is reached, the microprocessor pops some registers off of the stack and returns from the CLR subroutine. Otherwise, the microprocessor executes the INIT loop again.

At this point, you may wonder why we don't simply address the ten LED modules (one address per module) and, then, turn off all of the cursors in each module. This would certainly be possible and the INIT loop would only have to be executed 10 times, rather than 40 times. The problem is, a blank has to be written into *each* character position, since there is no way that we can "blank" four characters at once. Thus, we end up turning off the cursors in each module four times (by using the value 80H) which wastes a little time, but there is no simple way of getting around this. The loop that saves the blank characters has to be executed 40 times—once for each character position.

In Examples 6-7 and 6-8, we have assumed that the memory address of the first character in the 40-character string that is to be displayed has been loaded into one or more registers or memory loca-

Example 6-5. Initializing the 40-Character Display.

1. Using an 8085 microprocessor:

CLR	PUSH PSW	;Save A & flags
	PUSH B	;Save B & C
	MVI B,00	;Initial position
INIT	MOV A,B	;Get the position
	OUT 5EH	;Output to display
	MVI A,80H	;Cursors off
	OUT 5FH	;Output to display
	MVI A,' '	;A = ASCII space
	OUT 5FH	;Output to display
	INR B	;Increment position
	MOV A,B	;Get position in A
	CPI 28H	;Done all 40?
	JNZ INIT	;No, do another
	POP B	;Restore B & C
	POP PSW	;Restore A and flags
	RET	

2. Using a Z-80 microprocessor:

CLR	PUSH AF	;Save A & flags
	PUSH BC	;Save B & C
	LD B,00H	;Initial position
INIT	LD A,B	;Get the position
	OUT (5EH),A	;Output to display
	LD A,80H	;Cursors off
	OUT (5FH),A	;Output to display
	LD A,' '	;A = ASCII space
	OUT (5FH),A	;Output to display
	INC B	;Increment position
	LD A,B	;Get position in A
	CP 28H	;Done all 40?
	JR NZ,INIT	;No, do another
	POP BC	;Restore B & C
	POP AF	;Restore A and flags
	RET	

tions. This must be done *before* the OUT40 subroutine is called. In the subroutine, all the microprocessor has to do is set up a register

Example 6-6. Initializing the 40-Character Display.

1. Using a 6502 microprocessor:

CLR	PHA	;Save A
	PHP	;Save flags
	LDX #$00	;Initial position
INIT	STX $C05E	;Output the position
	LDA #$80	;Cursors off
	STA $C05F	;Output to display
	LDA #$20	;ASCII space
	STA $C05F	;Output to display
	INX	;Increment position
	CPX #$28	;Done all 40?
	BNE INIT	;No, do another
	PLP	;Restore the flags
	PLA	;Restore A
	RTS	

2. Using a 6800 microprocessor:

CLR	PSH A	;Save A
	PSH B	;Save B
	CLR B	;Initial position
INIT	STA B $C05E	;Output the position
	LDA A #$80	;Cursors off
	STA A $C05F	;Output to display
	LDA A #$20	;ASCII space
	STA A $C05F	;Output to display
	INC B	;Increment position
	CMP B #$28	;Done all 40?
	BNE INIT	;No, do another
	PUL B	;Restore B
	PUL A	;Restore A
	RTS	

with the character position, output the character position, read a value from memory and output it, increment the memory address, and increment the character position. If the character position is less than 28H, the LOOP1 loop is executed again. Otherwise, the microprocessor returns from the subroutine.

This software can be a lot "fancier" so that if a 10-character message needs to be displayed, only 10 characters are output. If this is done, you must blank the entire display (all 40 characters) before the message is output, so that part of the previous message is not left in the display. If the display contains the message "THE TIME IS 2:15 PM" and you output the new message "ALL LIGHTS OFF" without blanking the display, the message "ALL LIGHTS OFF15 PM" will be displayed.

Example 6-7. Displaying a String of 40 Characters on the Enlode 214 Display System.

1. Using an 8085 microprocessor:

```
              .
              .
              .
           PUSH H           ;Save H & L
           LXI H,STRING     ;Get the string address
           CALL OUT40       ;Display 40 characters
           POP H            ;Restore H & L
              .
              .
              .
OUT40:     PUSH B           ;Save B & C
           PUSH PSW         ;Save A & flags
           MVI B,00H        ;Initial position
LOOP1:     MOV A,B          ;Get the position
           OUT 5EH          ;Output to display
           MOV A,M          ;Get a character
           OUT 5FH          ;Output to display
           INX H            ;Increment pointer
           INR B            ;Increment position
           MOV A,B          ;Get position into A
           CPI 28H          ;Done all 40?
           JNZ LOOP1        ;No, output another
           POP PSW          ;Restore A & flags
           POP B            ;Restore B & C
           RET
```

Example 6-7. Cont.

2. Using a Z-80 microprocessor:

⋮

```
                PUSH HL             ;Save H & L
                LD HL,STRING        ;Get the string address
                CALL OUT40          ;Display 40 characters
                POP HL              ;Restore H & L
```
⋮
```
OUT40           PUSH BC             ;Save B & C
                PUSH AF             ;Save A & flags
                LD B,00H            ;Initial position
LOOP1           LD A,B              ;Get the position
                OUT (5EH),A         ;Output to display
                LD A,(HL)           ;Get a character
                OUT (5FH),A         ;Output to display
                INC HL              ;Increment pointer
                INC B               ;Increment position
                LD A,B              ;Get position into A
                CP 28H              ;Done all 40?
                JR NZ,LOOP1         ;No, output another
                POP AF              ;Restore A & flags
                POP BC              ;Restore B & C
                RET
```

THE ANALOG DEVICES RTI-1260

In Chapters 3 and 4, interfacing techniques for digital-to-analog and analog-to-digital converters were discussed. In some cases, you could save time and money by buying a DAC or ADC card rather than designing and building your own. This was particularly true if you needed a number of analog inputs in your system.

The RTI-1260 analog input board shown in Fig. 6-7 has 16 analog inputs which are multiplexed, one at a time, to the input of a 12-bit ADC. By adding some additional chips, the card can be used with up

Example 6-8. Displaying a String of 40 Characters on the Enlode 214 Display System.

1. Using a 6502 microprocessor:

    .
    .
    .

```
              PHA              ;Save A
              LDA #<STRING     ;Get MSBs of address
              STA $41          ;Save in page 0
              LDA #>STRING     ;Get LSBs of address
              STA $40          ;Save in page 0
              JSR OUT40
              PLA
              .
              .
              .

OUT40         STY $42          ;Save Y
              LDY #$00         ;Initial position
LOOP1         STY $C05E        ;Output the position
              LDA ($40),Y      ;Get a character
              STA $C05F        ;Output it
              INY              ;Increment position and index
              CPY #$28         ;Done all 40?
              BNE LOOP1        ;No, do another
              LDY $42          ;Restore Y
              RTS
```

2. Using a 6800 microprocessor:

    .
    .
    .

```
              STX $40          ;Save index register
              LDX STRING       ;Get address of string
              JSR OUT40
              LDX $40
              .
              .
              .
```

Example 6-8. Cont.

OUT40	PSH B	;Save B
	CLR B	;Clear B
LOOP1	STA B $C05E	;Output the position
	LDA A 0,X	;Get a character
	STA A $C05F	;Output it
	INX	
	INC B	;Increment position and index
	CMP B #$28	;Done all 40?
	BNE LOOP1	;No, do another
	PUL B	;Restore B
	RTS	

Figure 6-7.

The RTI-1260 and RTI-1262 boards are 12-bit analog input and analog output boards compatible with STD bus computers. *(Courtesy Analog Devices)*

to 32 analog signals. The analog signals that are wired to the inputs of the RTI-1260 can be either single-ended, pseudo-differential, or full differential. (These terms deal with the type of ground available from the signal source.) It is beyond the scope of this book to discuss different grounding techniques and problems, but some basic information about these three input configurations (from Analog Devices) has been included in Appendix B.

The analog output of the multiplexer is wired to a programmable gain amplifier, where the gain can be set to between 1 V/V and 1000 V/V, depending on the value of a programming resistor (R_g). By using the equation,

$$G = 1 + \frac{20 \text{ kilohms}}{R_g}$$

the value of the programming resistor can be determined for any gain. Thus, for a gain of 10, a 2222-ohm resistor would be used. This resistor should be a metal-film resistor, with a very low-temperature coefficient ($\pm$ 25 ppm/°C).

The output of the programmable gain amplifier is wired to a sample-and-hold, which "memorizes" the analog input voltage while the ADC is performing a conversion. This prevents the analog voltage from varying while the conversion is taking place.

The output of the sample-and-hold is wired to the 12-bit ADC, which can be used to convert analog voltages in the range of 0–10 V, or $\pm$ 10 V. The digital outputs of the ADC can represent the analog input voltage using either binary, offset binary, or 2's complement numbers. The differences between these three types of binary numbers are summarized in Table 6-5.

Table 6-5. Binary, Offset Binary, and 2's Complement Numbers

Voltage	Binary	Offset Binary	2's Complement
10 V	111111111111	111111111111	011111111111
0 V	000000000000	100000000000	000000000000
−10 V	*	000000000000	100000000000

*Binary numbers are only available for the 0–10 V range.

As always, the board has to be jumpered to respond to an address. In the case of the RTI-1260 board, it can only be interfaced to an STD bus microcomputer as a memory-mapped I/O device, but it does have the capability of using the memory-expansion signal, MEMEX. Thus, the RTI-1260 could be strapped for a portion of one 65K block of memory, and the microcomputer read/write and read-only memory could be strapped for the other 65K block. Of course, this means that the memory cards have to include the MEMEX signal in their decoders, in order for this to work.

The RTI-1260 board looks like a block of three memory locations, where the jumpered addresses for the card are XXXX 1111 XXXX 1011, XXXX 1111 XXXX 1100, and XXXX 1111 XXXX 1101. The Xs in these three addresses represent either 1s or 0s, depending on which jumpers are used. Thus, there are jumpers for A15, A14, A13, A12, and A7, A6, A5 along with A4. The other eight address bits are fixed. The board comes from the factory already jumpered for addresses FFFB, FFFC, and FFFD. If you want to use this card with most 6800-type CPU cards, you will have to change these jumpers.

The relationship between the address jumpers and the addresses selected can be seen in Table 6-6. What jumpers would be required for the card to respond to addresses 8FC0, 8FC1, and 8FC2? It isn't possible to use these addresses on the RTI-1260 board, because the least-significant digit of the hex addresses must be B, C, and D.

Three decoded memory addresses are used by the RTI-1260 input board so that the microcomputer can output a multiplexer address

Table 6-6. Address Jumpers for Analog Devices' RTI-1260 ADC Board

Base Address	X Digit Jumpers				Y Digit Jumpers			
Hex Digit	5–6	13–14	11–12	3–4	7–8	15–16	9–10	1–2
0	*	*	*	*	*	*	*	*
1	*	*	*		*	*	*	
2	*	*		*	*	*		*
3	*	*			*	*		
4	*		*	*	*		*	*
5	*		*		*		*	
6	*			*	*			*
7	*				*			
8		*	*	*		*	*	*
9		*	*			*	*	
A		*		*		*		*
B		*				*		
C			*	*			*	*
D			*				*	
E				*				*
F								
Address Bit	A15	A14	A13	A12	A7	A6	A5	A4

Note: Install jumpers between posts indicated by asterisk for address digit desired.

Courtesy Analog Devices

and start the ADC, and, then, when the conversion is completed, the 12-bit ADC value can be input. The assignment of the three memory addresses to these ports is shown in Fig. 6-8.

BYTE ADDRESS	DATA FORMAT							FUNCTION	OPERATION
	D7	D6	D5	D4	D3	D2	D1	D0		
XFYB	ϕ	ϕ	ϕ	M_4	M_3	M_2	M_1	M_0	MUX ADDR/ CONV	WRITE
XFYC	B_7	B_6	B_5	B_4	B_3	B_2	B_1	LSB	A/D DATA LO	READ
XFYD	BUSY	0/MSB	0/MSB	0/MSB	MSB/MSB	B_{10}	B_9	B_8	A/D DATA HI	READ

NOTES: 1. X AND Y ARE USER SELECTABLE.
        2. BITS SHOWN AS ▭ HAVE THE UPPER VALUE FOR UNIPOLAR CODES AND LOWER VALUE FOR 2's COMPLEMENT.
        3. THE SYMBOL ϕ MEANS THE BIT IS IGNORED.
        4. BUSY BIT EQUALS "1" DURING CONVERSIONS AND "0" WHEN DONE.

FIGURE 6-8

Structure of the three RTI-1260 memory locations. *(Courtesy Analog Devices)*

The address of the channel whose analog voltage is to be converted is output to port XFYB, where X and Y are user selectable. The number of multiplexer channels will be determined by how the inputs are strapped (single-ended, pseudo-differential, or differential), along with whether an extra set of analog multiplexer chips has been plugged into the board. In the single-ended and pseudo-differential modes of operation, there will be 16 inputs, using multiplexer channel numbers 0–15 (decimal; 0–F hex). If the extra multiplexer chips are added, the channel numbers will be between 0 and 31. If the differential mode of operation is used, the number of channels is cut in half so there can be either 8 or 16 differential channels, depending on whether or not the extra multiplexer chips have been plugged into the board.

The same pulse that is used to latch the multiplexer channel number is also used to start the ADC. While the ADC is converting, bit D7 of port XFYD will be a logic 1. Once the conversion has been completed, this bit will go to a logic 0. This is the flag bit that the software must monitor. Once this bit is a logic 0, the 12 bits of ADC data can be input, using ports XFYC and XFYD. Note that the states of the bits in port XFYD depend on whether or not 2's complement numbers are to be input from the converter.

How long does a conversion take? This depends on the gain of the programmable gain amplifier. The higher the gain, the longer it takes for the output of the programmable amplifier to settle down to its final value. A delay circuit has been incorporated into the RTI-1260 input board so that the conversion is not started until the amplifier has settled down. For gains between 1 and 1000, the delay ranges between 13 μsec and 85 μsec. Thus, if the gain is 1, it will take 13 μsec for the amplifier to settle down and 25 μsec for the ADC to perform the conversion. Thus, over 25,000 conversions/second can be performed. At a gain of 1000, 85 + 25 μsec are required, for a conversion rate of 9090 conversions/second. *After the user determines the gain for the RTI-1260, the appropriate capacitor has to be selected for the delay circuit.* The User's Manual for the RTI-1260 board contains a table of delays caused by readily available capacitors.

In order to use this board in a system, a number of jumpers have to be added or changed. These include:

1. Address jumpers (8 jumpers).
2. A/D input range (0–10 V or ± 10 V; 2 jumpers).
3. Multiplexer configuration (single-ended, pseudo-differential, or differential; 3 jumpers).
4. A/D code (binary, offset binary, or 2's complement; 2 jumpers).

In addition, the user has to add a gain resistor to the board, along with the appropriate delay capacitor.

The configuration of the board from the factory is as follows:

1. 16 single-ended channels.
2. A gain of one (1).
3. Conversion delay of 15 μsec.
4. A/D input range of ± 10 V, offset binary.
5. Memory addresses FFFB, FFFC, and FFFD.

The software for this board can be either very simple or very complex, depending on the number of channels being used. A simple general-purpose ADC subroutine is listed in Example 6-9. Before this subroutine is called, the A register or accumulator must contain a multiplexer channel number. When the subroutine is called, this value is output to the RTI-1260 board. By doing this, the channel is selected and the ADC is started. If this subroutine is called with an

Example 6-9. A Simple ADC Subroutine for the RTI-1260 Interface Board.

1. Using the 8085 microprocessor:

ADCSUB:	STA 0FFFBH	;Output mux. address
ADCS1:	LDA 0FFFDH	;Get the ADC flag
	ORA A	;Set/clear flags
	JM ADCS1	;Flag is 1, loop
	LHLD 0FFFCH	;Get 12-bit value
	RET	

2. Using the 6502 microprocessor:

ADCSUB	STA $FFFB	;Output mux. address
ACDS1	BIT $FFFD	;Test ADC flag
	BMI ADCS1	;Wait for a 0
	LDA $FFFC	;Get the 8 LSBs
	TAX	;Save them in X
	LDA $FFFD	;Get 4 MSBs
	RTS	

3. Using the Z-80 microprocessor:

ADCSUB	LD (FFFBH),A	;Output mux. address
ADCS1	LD A,(FFFDH)	;Get the ADC flag
	OR A	;Set/clear flags
	JP M,ADCS1	;Flag is 1, loop
	LD HL,(FFFCH)	;Get 12-bit value
	RET	

4. Using the 6800 microprocessor:

ADCSUB	STA A $FFFB	;Output mux. address
ADCS1	TST $FFFD	;Test ADC flag
	BMI ADCS1	;Wait for a 0
	LDX $FFFC	;Get 12-bit value
	RTS	

invalid channel number, the ADC will still perform a conversion but the result will be meaningless. This subroutine could be improved by first executing a compare-immediate instruction—where the data byte of the instruction is the number of the highest channel. If the A register contains a number greater than this when the subroutine is called, an error subroutine might be called, or some other action might be taken.

Once the channel number has been output, the microprocessor waits for the ADC flag to indicate that the conversion has been performed. Once it has been performed, the 12-bit value is input and, then, the microprocessor returns from the ADCSUB subroutine. Unfortunately, the RTI-1260 input board cannot be used to interrupt the microprocessor when a conversion has been performed, so the microprocessor must always be programmed to wait for the ADC flag to go from a logic 1 to a logic 0.

In many data-acquisition applications, the ADC software is not this simple. For example, assume that you have to acquire a different number of data points from different channels, all at different rates. You might need to acquire 500 points with 45 msec between points from one channel, 2360 points at the rate of 1 point/sec from another channel, and 234 points from another channel at the rate of 1 point every 6.534 seconds. With such different data-acquisition rates, the software starts to get complex.

One way to solve this type of problem is to interrupt the microprocessor every 1 msec and, each time the interrupt occurs, the microprocessor decrements a count associated with each channel. Once this count is 0, the microprocessor knows to acquire data from this channel. Thus, for the previous three channels, the counts would be 45, 1000, and 6534, respectively. Once the count has been decremented to 0, not only must a conversion be performed, but the memory locations that contain the count must also be reinitialized. Thus, two memory locations can be used to store the count that is being decremented, and two additional memory locations are used to store the original count.

In addition, the microprocessor has to store the number of values to be acquired from each channel, the memory address where the data is to be stored, and the multiplexer channel number associated with the data being acquired. Thus, for each channel, there is a 9-byte channel control block. The organization of this channel control block is shown in Fig. 6-9.

The first byte in the channel control block (CCB) contains either the multiplexer channel number, or an inactive channel indicator (40H). Ordinarily, the channel number in this byte is output to the RTI-1260 board as the multiplexer channel number. However, once all of the

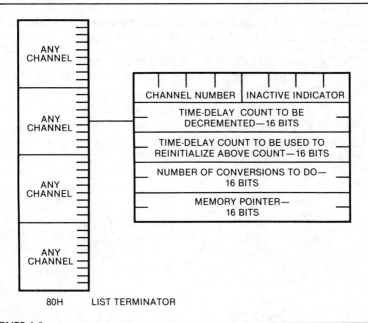

FIGURE 6-9
Structure of channel control block for the RTI-1260 input board.

data values for this channel have been acquired and stored in memory, the microprocessor stores a 40H (an invalid channel number) in this memory location. When the microprocessor reads this value from memory, it knows that all of the data values have been acquired for this channel, so it skips the next eight memory locations. After the channel number, a 16-bit time-delay count is stored in memory. This is the number of interrupts that must occur before a conversion for this channel is performed. If an interrupt is generated every 2 msec, a maximum delay of 131 seconds is possible. When this count is decremented to 0, the original value (stored in the next two consecutive memory locations) is used to reinitialize the count. The next two memory locations contain the number of conversions to be performed for this channel. When this value is decremented to 0, the memory location that contains the channel number is loaded with 40H to disable this channel. The last two bytes of the CCB contain the memory pointer used to store the ADC data in memory.

The software that uses these CCBs is listed in Examples 6-10 through 6-12. We have not tried this software so it may still have a few bugs in it. As you can see, the software is pretty complex. This is

Example 6-10. A Multichannel General-Purpose ADC Interrupt Service Subroutine for the 8085 Microprocessor.

MUXADC:	PUSH PSW	;Save A and the flags
	PUSH B	;Save registers B and C
	PUSH D	;Save registers D and E
	PUSH H	;Save registers H and L
	LXI H,TABLE	;Get the CCB table address
ENDCHK:	PUSH H	;Save the table address
	MOV A,M	;Get the 1st byte from a CCB
	CPI 40H	;Is this channel active?
	JZ SKIPIT	;No, so skip this CCB
	CPI 80H	;At the end of the table?
	JNZ MORE	;No, so process the CCB
	POP H	;Yes, clean up the stack
	POP H	;and restore the registers
	POP D	
	POP B	
	POP PSW	
	EI	;Enable the interrupt
	RET	;Return to the interrupted task
MORE:	STA 0FFFCH	;Output multiplexer channel, start the ADC
	INX H	;Now address the 16-bit delay value
	MOV D,H	;Move the high part of the address to D
	MOV E,L	;Move the low part to E
	INR M	;Increment the 2's complement count
	JNZ SKIPIT	;Nonzero LSBs, not time for a conversion
	INX H	;LSBs are zero, how about MSBs?
	INR M	
	JNZ SKIPIT	;Nonzero MSBs, don't get an ADC value
	INX H	;Count is 0, reinitialize it

Example 6-10. Cont.

	MOV A,M	;Get the LSBs of original count
	STAX D	;Store it
	INX H	;Increment both
	INX D	;addresses
	MOV A,M	;Get the MSBs of the original count
	STAX D	;and store it
	INX H	;Now address the number of points
	INR M	;Has the 16-bit point count
	JNZ DOADC	;been incremented to 0? If not,
	INX H	;do an ADC and store the value.
	INR M	
	JNZ DOADC1	
	POP H	;Count is 0, so disable this
	MVI M,40H	;channel
	JMP SKIP1	;Access the next channel
DOADC:	INX H	;Get memory address from table
DOADC1:	INX H	
	MOV E,M	;LSBs of address into register E
	INX H	;Now address the MSBs
	MOV D,M	;MSBs of address into register D
WAIT:	LDA 0FFFDH	;Get the ADC flag
	ORA A	;Set the 8085's flags
	JM WAIT	;Wait for a logic 0 in D7
	STAX D	;Save the MSBs of the ADC value
	INX D	;Increment the list address
	LDA 0FFFCH	;Get the LSBs
	STAX D	;Save them in memory
	INX D	;Increment the list address
	MOV M,D	;Save the address MSBs back in the CCB
	DCX H	;Decrement the CCB table address

Example 6-10. Cont.

	MOV M,E	;Save the LSBs of the address
SKIPIT:	POP H	;Get the address of 1st byte in CCB
SKIP1:	LXI D,0009H	;Get an offset into D and E
	DAD D	;Add the offset to the address in HL
	JMP ENDCHK	;Test the next CCB (if there is one)

Example 6-11. A Multichannel General-Purpose ADC Interrupt Service Subroutine for the Z-80 Microcomputer.

MUXADC	PUSH IX	;Save the IX register
	EX AF,AF'	;Swap A and the flags
	EXX	;Swap B, C, D, E, H and L
	LD IX,TABLE	;Get the table address into IX
ENDCHK	LD A,(IX+0)	;Get the 1st byte from a CCB
	CP 40H	;Has this channel been deactivated?
	JP Z,SKIPIT	;Yes, so test the next CCB
	CP 80H	;At the end of the table of CCBs?
	JP NZ,MORE	;No, and this channel is active
	EXX	;At the end of the table, restore B-L
	EX AF,AF'	;Restore A and the flags
	POP IX	;Restore the IX index register
	RETI	;Return from the interrupt
MORE	LD (FFFB),A	;Output the mux. address, start the ADC
	INC (IX+1)	;Increment the time delay LSBs
	JP NZ,SKIPIT	;Not 0 yet, so don't get an ADC value
	INC (IX+2)	;Increment the time delay MSBs.
	JP NZ,SKIPIT	;Not 0 either, so check the next CCB

Example 6-11. Cont.

	LD A,(IX+3)	;Count is 0, so restore the count
	LD (IX+1),A	;First restore the LSBs
	LD A,(IX+4)	;Then restore the MSBs
	LD (IX+2),A	
	INC (IX+5)	;Now see if the point count is 0
	JP NZ,DOADC	;If nonzero, get an ADC value
	INC (IX+6)	;LSBs were 0, how about the MSBs?
	JP NZ,DOADC	;No, the MSBs are nonzero
	LD A,40H	;Point count is 0, deactivate
	LD (IX),A	;this channel
	JP SKIPIT	;Now try the next CCB
DOADC	LD L,(IX+7)	;Get address LSBs from CCB into L
	LD H,(IX+8)	;Get address MSBs from CCB into H
WAIT	LD A,(FFFDH)	;Get the ADC flag (and the MSBs)
	OR A	;Set the Z-80's flags
	JP NZ,WAIT	;Wait for the flag to be 0
	LD (HL),A	;Save the MSBs
	INC HL	;Increment the memory address
	LD A,(FFFCH)	;Get the LSBs of the ADC value
	LD (HL),A	;Save them in memory
	INC HL	;Increment the memory address
	LD (IX+7),L	;Save the new address back in the CCB
	LD (IX+8),H	
SKIPIT	LD DE,0009H	;Get an offset into D and E
	ADD IX,DE	;Add to IX, result in IX
	JMP ENDCHK	;Now check the next CCB

Example 6-12. A Multichannel General-Purpose ADC Interrupt Service Subroutine for the 6502 Microprocessor.

MUXADC	PHA	;Save A on the stack
	STX $38	;Save X on page 0
	LDX #$00	;Load an index of 0
ENDCHK	LDA LIST,X	;Get the first byte from the CCB
	CMP #$40	;Is the channel active?
	BEQ SKIPIT	;No, access the next CCB
	CMP #$80	;At the end of the table?
	BNE MORE	;No, must be a channel number
	LDX $38	;Done all channels, restore X
	PLA	;Restore A
	RTI	;Return from the interrupt
MORE	STA $CFFC	;Output the channel number, start the ADC
	INC LIST+1,X	;Increment the LSB time delay value
	BNE SKIPIT	;Not 0, so don't digitize this channel
	INC LIST+2,X	;0, so increment the MSBs
	BNE SKIPIT	;Nonzero result, so skip this channel
	LDA LIST+3,X	;0 result, get the original LSBs
	STA LIST+1,X	;Store in the LSB count location
	LDA LIST+4,X	;Get the original MSBs
	STA LIST+2,X	;Store in the MSB count location
	INC LIST+5,X	;Increment the number of points
	BNE DOADC	;to be acquired
	INC LIST+6,X	;LSBs are 0, so increment the MSBs
	BNE DOADC	;Not 0, so get the ADC data
	LDA #$40	;Point count is 0, so disable this

Example 6-12. Cont.

	STA LIST,X	;channel (store a $40)
	JMP SKIPIT	;Access another channel
DOADC	BIT $CFFD	;Check the ADC's flag
	BMI DOADC	;Flag is a 1, so wait
	LDA $CFFC	;Get the LSBs
	STA (LIST+7,X)	;Store the LSBs
	INX	;Increment the index
	LDA $CFFD	;Get the MSBs
	STA (LIST+7,X)	;Store the MSBs
	DEX	;Decrement the index
	CLC	;Clear the carry
	LDA #$02	;Add 2 to the address in the table
	ADC LIST+7,X	;Add the table address
	STA LIST+7,X	;Save the result
	LDA #$00	
	ADC LIST+8,X	;Add the MSBs of the address
	STA LIST+8,X	;Save the result
SKIPIT	CLC	;Clear the carry
	TXA	;Transfer the index to A
	ADC #$09	;Add the number of bytes in a CCB
	TAX	;Put the result back in X
	JMP ENDCHK	;Access the next CCB

due to the fact that the microprocessor has to be able to access a number of memory locations. The 8085 software is particularly complex, because the 8085 processor does not have an indexed addressing mode.

This group of examples (Examples 6-10 through 6-12) is the only group that does not contain any 6800 software. Unfortunately, there is no easy way to get the memory pointer (index) from one of the CCBs into the X index register, use it to address memory when the ADC values are stored in memory, increment it, and save it back in the CCB. When Motorola designed the 6809 integrated circuit, they solved this problem by adding another index register (Y) along with an indexed indirect addressing mode.

Basically, the microprocessor will be interrupted every one or two milliseconds. When this happens, registers are saved on the stack and, then, the microprocessor starts to access the CCBs. If the value read from the table is an 80H, the end of the table has been reached, so the microprocessor can return from the interrupt-service subroutine. If the value read is a 40H, the microprocessor jumps down to SKIPIT, where 9 is added to the table memory address or index. If the value read is neither of these two values, it must be a valid channel number so it is output to the multiplexer and a conversion is begun at MORE.

After the multiplexer value is output, a 16-bit count in memory is decremented. This is the number of interrupts that must occur before a value is acquired. If this value is decremented to a nonzero value, the microprocessor jumps to SKIPIT. If this count is decremented to 0, the original and unaltered count has to be read from the next two consecutive memory locations and is stored in the two memory locations that are decremented. Since the count has now been decremented to 0, the word count or conversion count has to be decremented. This means that a value will be acquired and stored in memory. If the word count is decremented to 0, the channel has to be disabled, because all of the values for this channel have been acquired.

By the time the microprocessor reaches DOADC, it is ready to input a 12-bit ADC value and store it in memory—which is what it does. The memory address is incremented (or 2 is added to it) and it is saved back in the CCB. Finally, at SKIPIT, 9 is added to the starting address of this particular CCB so that the microprocessor can access the next CCB.

A few final notes. At MORE, the multiplexer value was output and the ADC started, even *before* we knew whether the time-delay count would be decremented to 0. This was done to save time and software. Ordinarily, the multiplexer value would be output once the time-delay count is decremented to 0. However, by starting the converter early, we will spend less time waiting for the flag to go to a logic 0 at DOADC or WAIT. Thus, the ADC is converting at the same time that the microprocessor is executing software.

Since interrupts are used, it is extremely important that you realize how long it will take the microprocessor to execute this software for one channel. On a 6.144-MHz 8085 microprocessor (320-nsec cycle time), between 130–150 μsec are required *per channel*. Thus, if all 16 channels have to be serviced (there are 16 CCBs in the table), an interrupt had better not occur every 1 msec. In fact, an interrupt better not occur every 2 msec either. The problem, of course, is that the interrupt will interrupt the execution of the interrupt-service subroutine. Of course, if you know that only 5 channels will ever be active at once, a 1-msec interrupt will work just fine. However, when writing software, always assume the worst. Thus, it would be best if the interrupt only occurred every 5 msec. With a 16-bit time-delay count, delays of up to 327 seconds can be generated.

Finally, because of the way in which the software was written, the CCBs don't have to be stored consecutively (channel 0 first, followed by 1, 2, etc.). Depending on your application, you could have the CCB for channel 9 first, followed by the CCBs for channels 15, 3, 6, 2, 10, and 11.

THE ATEC 710 THUMBWHEEL SWITCH INTERFACE

The Atec 710 thumbwheel switch interface (Fig. 6-10) consists of an interface that plugs into an STD bus system, along with up to 16 thumbwheel switches. The switches can be mounted on the interface board or on a remote control panel. If the switches are mounted on a control panel, a flat cable is used to interconnect the switches and the interface board.

In most cases, BCD thumbwheel switches will be used with this board. The BCD numbering system has 10 states, using 4 binary bits, that are used to represent the numbers 0–9. There are 6 illegal BCD numbers, represented by the binary numbers 1010, 1011, 1100, 1101, 1110, and 1111. Since only 4 bits are required for each switch, each three-state memory-mapped input port is used to interface two switches to the data bus. The format of the data value from each port is shown in Table 6-7.

Because there can be a maximum of 16 switches (two per port), this board uses eight consecutive memory addresses. *Also, this board can only be used as a memory-mapped I/O device.* Thus, this board looks like eight consecutive read-only memory locations. No information can be output to the interface.

Like all of the other boards that have been discussed, the 710 switch interface has jumpers that determine the addresses that the board responds to. Ordinarily, the board is shipped already jumpered for addresses 6000 through 6007. The location of the jumpers is shown in Fig. 6-11 and the relationship between the jumpers and decoded addresses is summarized in Table 6-8.

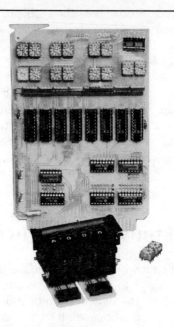

FIGURE 6-10

Using the Atec 710 interface with thumbwheel switches. *(Courtesy Atec, Inc.)*

Table 6-7. Data Format of the Input Ports on the Atec 710

Data Bit	D7	D6	D5	D4	D3	D2	D1	D0
Weight	8	4	2	1	8	4	2	1
Switch (Typical)	S1				S2			

Courtesy Atec, Inc.

From Table 6-8, what jumpers would be required for addresses 91C8 through 91CF? For these addresses, jumpers X1 to 4, X2 to 4, X3 to 3, X5 = 2, and X4 = 1 are needed. As with previous boards, the addresses used must be consecutive and, in this case, must be a multiple of 8. Thus, addresses such as C345 through C34C, or 0899 through 08A0, cannot be used.

Since this interface is just for reading values from thumbwheel switches, the software is relatively simple. As an example, let's input two 3-digit BCD values, add them, and save the result in memory. The software that does this is listed in Examples 6-13 and 6-14. There is nothing particularly unusual about this software. The only point to remember is that when the most-significant digit is read into the microcomputer, two digits are input. After these digits are added, the

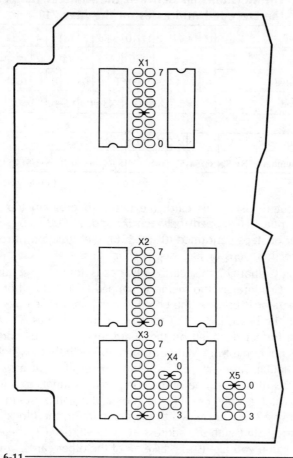

FIGURE 6-11
Location of address jumpers on the thumbwheel switch interface. *(Courtesy Atec, Inc.)*

result is ANDed with 0FH, so data bits D7–D4 are set to zero, and D3–D0 contain the most-significant digit in the 3-digit number. In other cases, the BCD values may have to be shifted left or right in order to align them before they are added or processed.

THE MATRIX 7911 STEPPER-MOTOR CONTROLLER

The Matrix 7911 Stepper-Motor Controller (SMC) card (Fig. 6-12) can be used to interface a variety of stepper motors to an STD bus

Table 6-8. The Relationship Between the Address Jumpers and the Decoded Addresses on the Atec 710

Address Bit	15	14	13	12	11	10	9	8	7	6	5	4	3	2	1	0
Hex Value	6				0				0				0–7			
Binary Value	0	1	1	0	0	0	0	0	0	0	0	0	*	*	*	
Jumper Pad	X1=3				X2=0				X3=0			X5=0	X4=0			

A0–A2 Address	0	1	2	3	4	5	6	7
Switch Number	S1/S2	S3/S4	S5/S6	S7/S8	S9/S10	S11/S12	S13/S14	S15/S16

Courtesy Atec, Inc.

microcomputer. Using this card, the user can program the rate at which the motor is stepped (between 50 and 30,000 steps/second), the number of steps performed (0–65,536), the direction of rotation, and whether full steps or half steps occur.

Surprisingly enough, this card only needs four consecutive accumulator I/O addresses (no memory-mapped I/O) to do all of this. A 6-bit comparator is used on the board to compare the settings of DIP switches on the board to the I/O address on address lines A7–A2. A1 and A0 are further decoded on the board for the various ports. The board does not come from Matrix Corporation with these switches set to any particular value so you will have to set them. If a switch is closed or ON, it will set one of the comparator's inputs to a logic 0, and if the switch is open or OFF, the comparator will be set to a logic 1. Since only A7–A2 are wired to the comparator, the block of four addresses must start with the address X0, X4, X8, or XC, where the X is any hex value you like (0–F). Like all of the other cards, you can't cross the multiple-of-4 boundary and use addresses 3E–41 or B5–B8. To set up the card for addresses A8–AB, the switch settings in Table 6-9 would be used.

What are the four ports on this card used for? This depends on whether a Read or Write operation is taking place. Using this card, you can output an 8-bit step rate, a 16-bit step count, and an 8-bit control word. The 8-bit step-rate value determines how frequently the stepper motor will be pulsed. This value can be between 50–12,750 steps/second or between 200–30,000 steps/second. The step-rate range that is used is determined by a jumper on the card. Since an 8-bit step rate can be written out to the card, we can control the step rate to within 50 steps/second on the low range ((12,750–50)/256) or 200 steps/second on the high range. Once the step rate is output, we

Example 6-13. Reading and Adding Thumbwheel Switch Values.

1. Using the 8085 microprocessor:

INPUT:	LDA 091C8H	;Get 2 digits
	MOV B,A	;and move them.
	LDA 091CAH	;Get 2 more
	ADD B	;Add them
	DAA	;Adjust result
	MOV L,A	;Save result
	PUSH PSW	;Save carry flag
	LDA 091C9H	;Get 2 more
	ANI 0FH	;Mask 1 out
	MOV B,A	;Save it
	LDA 091CBH	;Get 2 more
	ANI 0FH	;Mask 1 out
	POP PSW	;Get carry flag
	ADC B	;Add B & carry to A
	DAA	;Adjust result
	MOV H,A	;Save result
	SHLD ANS	;in memory
	RET	

2. Using the Z-80 microprocessor:

INPUT	LD A,(091C8H)	;Get 2 digits
	LD B,A	;and move them.
	LD A,(091CAH)	;Get 2 more
	ADD A,B	;Add them
	DAA	;Adjust result
	LD L,A	;Save result
	PUSH AF	;Save carry flag
	LD A,(091C9H)	;Get 2 more
	AND 0FH	;Mask 1 out
	LD B,A	;Save it
	LD A,(091CBH)	;Get 2 more
	AND 0FH	;Mask 1 out
	POP AF	;Get carry flag
	ADC A,B	;Add B & carry to A
	DAA	;Adjust result
	LD H,A	;Save result
	LD (ANS),HL	;in memory
	RET	

Example 6-14. Reading and Adding Thumbwheel Switch Values.

1. Using the 6502 microprocessor:

INPUT	CLC	;Clear the carry
	SEC	;Decimal mode
	LDA $91C8	;Get 2 digits
	ADC $91CA	;Add 2 more
	STA $40	;Save the result
	LDA $91C9	;Get 2 more
	AND #$0F	;Mask 1 out
	STA $41	;Save the digit
	LDA $91CB	;Get 2 more
	AND #$0F	;Mask 1 out
	ADC $41	;Add the other 1
	STA $41	;Save in page 0
	RTS	

2. Using the 6800 microprocessor:

INPUT	LDA A $91C8	;Get 2 digits
	ADD A $91CA	;Add 2 more
	DAA	;Decimal adjust
	STA A $41	;Save the result
	LDA A $91C9	;Get 2 more
	AND A #$0F	;Mask 1 out
	STA A $40	;Save the digit
	LDA A $91CB	;Get 2 more
	AND A #$0F	;Mask 1 out
	ADC A $40	;Add the other 1
	DAA	;Decimal adjust
	STA A $40	;Save in page 0
	RTS	

cannot read it back into the microprocessor. Regardless of the direction of rotation or the number of steps taken, the step rate will not be changed until a new value is output to the interface.

At first, it may seem unusual that the step rate can be controlled to 200 steps/second and within the range of 200–30,000 steps/second. You might expect that it could be controlled to within 116 steps/second ($\frac{30,000-200}{256}$). With 256 different 8-bit values, the upper step limit

FIGURE 6-12
The Matrix 7911 Stepper-Motor Controller. *(Courtesy Matrix Corp.)*

Table 6-9. Setting the 7911 SMC Card for Addresses A8–AB

Address	Switch	On/Off	Binary Value
A7	S1	Off	1 (8)
A6	S2	On	0 (4)
A5	S3	Off	1 (2)
A4	S4	On	0 (1)
A3	S5	Off	1 (8)
A2	S6	On	0 (4)
A1	None	-	0 (2)
A0	None	-	0 (1)

Courtesy Matrix Corp.

should be 256 × 200 steps/second, or 51,200 steps/second. The problem is that some of the circuitry on the interface cannot be used with step rates greater than 30,000 steps/second. Thus, in the range of 200–30,000 steps/second, you *can* control the step rate to within 200 steps/second. This also means that the 30,000 steps/second rate will be obtained when a value near 96H is used. *No step-rate values greater than this must be used when the high range of step rates (200–30,000) is jumpered on the board.*

In order to output a 16-bit step count, the interface has to use two 8-bit output ports—which it does. This means that the stepper motor can be stepped up to 65,535 steps in either a clockwise or counterclockwise direction. Once the step count is output, it cannot be read back into the microcomputer.

The fourth output is the control register. This port is used to control the direction of rotation, to enable and disable interrupts, to enable and disable the motor-control logic, and to cause either half steps or full steps to be used. These four output ports, along with the function of the input ports with the same addresses, are summarized in Table 6-10.

Table 6-10. Port Address and Port Functions for the 7911 SMC Card

A1	A0	Write Operation	Read Operation
0	0	Write to control register	Read the status flags
0	1	Write step count MSB	Read the status flags
1	0	Write step count LSB	Read the status flags
1	1	Write the step rate*	Clear the status flags*

*Both of these operations clear the status flags.

As you might expect, the status flags indicated in this table tell us whether or not the motor is moving, whether or not the step count has been decremented to 0, or whether an error has occurred. The function of each bit in the control register is shown in Table 6-11.

Note that many of the bits in both the status flag register and the control register don't even have to be used. However, these bits are brought out to the stepper-motor connector for your use.

Bit D7 of the control register determines whether or not an interrupt will occur. The interrupt logic on the SMC card can be wired to either the NMIRQ* or INTRQ* interrupt inputs. What condition will cause an interrupt? An interrupt can occur when the step-count register has been decremented to 0, when one step has occurred, when there is a ramp control error (which is generated by the logic on the interface), or when the stepper is stationary (not busy). Once the interrupt occurs, you would have to poll these flags to determine which condition caused the interrupt. Also, if these flags are wired to the INTRQ* line, the interrupt would have to be enabled.

The motor enable/disable bit in the control register can be used to

Table 6-11. Status Flags and Control Bits in the Control Register

Data Bit	Status	Read	Write
D7	0	Step control not 0	Disable interrupts
	1	Step control = 0	Enable interrupts
D6	0		Motor logic enabled
	1	Motor step occurred	Motor logic disabled
D5	0		CCW phase rotation
	1	Ramp control error	CW phase rotation
D4	0	Stepper busy	Full step sequence
	1	Stepper not busy	Half step sequence
D3	0	Aux. input 4=0	*Damping pulse disabled
	1	Aux. input 4=1	Damping pulse enabled
D2	0	Aux. input 3=0	Aux. output 3=0
	1	Aux. input 3=1	Aux. output 3=1
D1	0	Aux. input 2=0	Aux. output 2=0
	1	Aux. input 2=1	Aux. output 2=1
D0	0	Aux. input 1=0	Aux. output 1=0
	1	Aux. input 1=1	Aux. output 1=1

*The damping enable jumper J9 must be installed 1−2 for this function. Otherwise, a Write will control auxiliary output 4 in same manner as D0−D2.

Courtesy Matrix Corp.

stop the stepper motor and the "direction" bit will determine the direction of rotation for the stepper motor. The full-step/half-step bit controls whether or not a full or half step is taken by the motor. As an example, if a motor has 400 steps/revolution, then 800 half steps could be taken. Steps are also specified in degrees, where a full step might be 15° and a half step would be 7.5°. The damping pulse bit is used to generate a pulse in the opposite direction to the direction that the motor just moved—just before the stepper comes to rest.

At some point, it may seem unusual that there are flags to indicate whether or not the step counter has been decremented to 0 and whether or not the stepper motor is moving. At first glance, you would expect that these flags would change at the same time, because when the step count is decremented to 0, the motor should be stationary. However, the step count is usually decremented to 0 before the motor has reached its final position.

Let's look at a typical stepper-motor application to see why this is the case. Stepper motors are often used in numerically controlled milling machines and lathes, where the stepper motor either positions the piece of metal being worked on or the tool being used. Regardless of whether the metal or the tool is being moved, when the stepper motor first tries to move the object, there is a large resistance to overcome. Thus, the stepper should start out slowly and accelerate up to a set rate (the rate that *you* programmed into the step-rate register). Once the object gets near its final position, the stepper should slow down so that there isn't much momentum, and so that the final position isn't overshot or missed.

In a simple stepper-motor interface, where the rate at which *we* pulse the motor determines the stepping rate, the software needed to increase the stepping rate to a fixed value, maintain the fixed stepping rate, and to decrease the stepping rate would be pretty complex. Instead of making the user do this, Matrix Corporation used some special stepper-motor control chips on their board so that all the user has to do is output the step rate that is to be maintained and the number of steps that are to be taken. The interface takes care of increasing and decreasing the step rate.

Of course, the controller begins to increase the step rate to its final value as soon as you start the stepper motor moving, but how does the controller know when to decrease the stepping rate? *The stepping rate is decreased when the step counter is decremented to 0.* At first, it may look as if the stepping motor is stepped the number of times that you output and, then, some additional steps, but it is not. The special stepper-motor control chips have a "digital buffer" in them, so that when the step count has been decremented to 0, the counters in these chips will not have been decremented to 0, and these internal counters will indicate the number of steps that have taken place so far. It's not important that you understand how this works. However, if you want the stepper to step 5000 times, you output 5000 (decimal) to the controller and this is the number of steps that will be taken. The controller takes care of increasing the step rate up to the value that you output and, then, decreasing this value. Ordinarily, the stepper will start out at 300–500 steps/second and will finish up at this rate. Thus, the step rate will only start to decrease when the step count has been decremented to 0. At some later time, the stepper will finally stop moving.

So why bother with the step-counter flag? If you need to step the stepper 100,000 times, you can't do it by writing a 16-bit step count out to the interface. You could output a count of 50,000 for that number of steps to be taken and, then, output 50,000 again. To do this, you *would* monitor the step-counter flag, so that before the stepping rate has a chance to decrease, the motor continues to step at the rate that you output. If instead of monitoring this flag, you were to monitor the stepper busy/not busy flag, the stepper would step 50,000 times, but, toward the end, it would slow down and stop. At this time, a value of 50,000 would be output again, and the stepper would start to move again.

Example 6-15. Controlling a Stepper Motor with the 7911 SMC Card.

1. Using the 8085 microprocessor:

STOP	MVI A,40H	;Disable motor bit
	OUT A8H	;Stop the motor
	IN ABH	;Get and clear
	RET	;the flags
CW	MVI A,20H	;Rotation bit
	OUT A8H	;Output it
	JMP STEP	;Now step
CCW	XRA A	;A = 00000000
	OUT A8H	;Output it
STEP	LHLD STPCNT	;Get the step count
	MOV A,H	;Get the MSBs
	OUT A9H	;Output MSBs
	MOV A,L	;Get the LSBs
	OUT AAH	;Output the LSBs
	LDA STEPRT	;Get the step rate
	OUT ABH	;Output it
WAIT	IN A8H	;Get the flags
	ANI 10H	;Mask all but D4
	JZ WAIT	;Wait for a 1
	RET	;All done

2. Using the Z-80 microprocessor:

STOP	LD A,40H	;Disable motor bit
	OUT (A8H),A	;Stop the motor
	IN A,(ABH)	;Get and clear
	RET	;the flags

Example 6-15. Cont.

CW	LD A,20H	;Rotation bit
	OUT (A8H),A	;Output it
	JP STEP	;Now step
CCW	XOR A	;A = 00000000
	OUT (A8H),A	;Output it
STEP	LD HL,(STPCNT)	;Get the step count
	LD A,H	;Get the MSBs
	OUT (A9H),A	;Output MSBs
	LD A,L	;Get the LSBs
	OUT (AAH),A	;Output the LSBs
	LD A,(STEPRT)	;Get the step rate
	OUT (ABH),A	;Output it
WAIT	IN A,(A8H)	;Get the flags
	AND A,10H	;Mask all but D4
	JR NZ,WAIT	;Wait for a 1
	RET	;All done

In Examples 6-15 and 6-16, some simple stepper-motor control software has been given. Note that we have assumed that no counts over 65,535 will be used, so the stepper busy/not busy flag is monitored. To keep the software simple, no interrupts have been used either. The SMC software consists of three subroutines that can be used to (1) stop the stepper motor, (2) pulse it in one direction, and (3) pulse it in the opposite direction. In the STOP subroutine, the microprocessor simply sets the motor-logic disable bit (D6), writes this value out to the interface card, and, then, clears the SMC's flags. Once these operations have been performed, the microprocessor returns from the subroutine.

In the CW subroutine, the microprocessor sets the direction-of-rotation bit to a 1 and, then, outputs this value. The microprocessor then branches or jumps to STEP where the step count is output, followed by the step rate. The same instruction that outputs the step rate also causes all of the flags on the SMC card to be cleared. This is done so that the flag from the previous movement command is cleared, which means that the microprocessor is monitoring the stepper busy/not busy flag based on its present operation. The only difference between the CCW and CW subroutines is that the CCW subroutine clears the direction-of-rotation bit.

Example 6-16. Controlling a Stepper Motor with the 7911 SMC Card.

1. Using the 6502 microprocessor:

STOP	LDA #$40	;Disable motor bit
	STA $C0A8	;Output it
	LDA $C0AB	;Get and clear
	RTS	;the flags
CW	LDA #$20	;Rotation bit
	STA $C0A8	;Output it
	JMP STEP	;Step the motor
CCW	LDA #$00	;A = 00000000
	STA $C0A8	;Output it
STEP	LDA $40	;Get MSBs of
	STA $C0A9	;the step count
	LDA $41	;Get LSBs of
	STA $C0AA	;the step count
	LDA $42	;Get the step rate
	STA $C0AB	;Output it
WAIT	LDA $C0A8	;Get the flags
	AND #$10	;D4 = 0?
	BEQ WAIT	;Yes, so wait
	RTS	;Flag is a 1, return

2. Using the 6800 microprocessor:

STOP	LDA A #$40	;Disable motor bit
	STA A $C0A8	;Output it
	LDA A $C0AB	;Get and clear
	RTS	;the flags
CW	LDA A #$20	;Rotation bit
	STA A $C0A8	;Output it
	BRA STEP	;Step the motor
CCW	CLR A	;A = 00000000
	STA A $C0A8	;Output it

Example 6-16. Cont.

```
        STEP    LDX $40         ;Get the step count
                STX $C0A9       ;Output it
                LDA A $42       ;Get the step rate
                STA A $C0AB     ;Output it
        WAIT    LDA A $C0A8     ;Get the flags
                AND A #$10      ;D4 = 0?
                BEQ WAIT        ;Yes, so wait
                RTS             ;flag is a 1, return
```

THE PRO-LOG 7304 DUAL UART CARD

There are probably more crt terminals interfaced to microcomputers than any other peripheral device. Usually the crt interface is based on a UART or USART; both are MOS/LSI chips. The UART, or Universal Asynchronous Receiver/Transmitter, was developed in the early 1970s as a general-purpose communications chip. Because of this, it takes a few extra chips to interface the UART to a microcomputer. A more modern chip that has more functions built into it and is easier to interface to a microcomputer is the USART, or Universal Synchronous/Asynchronous Receiver/Transmitter. Regardless of whether the 40-pin UART or the 28-pin USART is used, each performs the same basic functions, which include receiving information from the crt terminal keyboard and transmitting information to the logic that controls the crt screen.

Since crt terminals may be located 100 or more feet from the microcomputer, it is not very practical to output 8 bits of information for the screen logic, and have this information go to the terminal over eight or sixteen wires (8 signal, 8 ground). Likewise, it is not very practical for the keyboard to be interfaced to an input port using eight or sixteen wires. The cost and sheer bulk of all this wire makes this a very unattractive interfacing method.

The UART and USART get around this problem by using between three and six wires and using a *serial* data format. This means that one bit of information at a time is either sent to the terminal or received from the terminal. By sending only one bit of information at a time, only two wires are required (one for the signal and one for a ground). The cost of the wire is certainly less and the electronics needed to drive a single wire is a lot less costly than the electronics required to drive eight wires.

As you might expect, if only one bit of information is received or transmitted at a time, some type of protocol has to be established. *As far as we know, all UARTs and USARTs transmit the LSB first and expect to receive the LSB first.* This assumes that pin D0 of the UART and USART is wired to D0 of the data bus, etc. Another protocol that has to be established is how that information will be represented on the *serial data link*. There are two common standards—20 mA and RS-232C. If a 20-mA loop is used, a current of 20 mA will flow through the circuit for a logic 1 data bit, and no current will flow for a logic 0 data bit. If RS-232C is used, a logic 1 data bit is represented by a voltage between -5 V and -15 V, and a logic 0 data bit is represented by a voltage between 5 V and 15 V.

What does the UART or USART actually do? To transmit a character to the terminal, you simply output an 8-bit value to the interface chip and it converts the 8-bit parallel value to an 8-bit serial value. In this arrangement, one bit is transmitted at a time—one right behind the other. The UART or USART also receives a serial character from the terminal and converts it to an 8-bit parallel value that can be input. Thus, these devices are very fancy shift registers, where they perform a serial-to-parallel and parallel-to-serial conversion. They perform a number of additional functions also, but this is basically what they do.

One more protocol has to be established in order to use one of these sophisticated communications devices with a terminal. How fast will information be transmitted or received? The following are some commonly used asynchronous communication *bit rates*. They can be used with most terminals.

$$\begin{array}{l}
110 \text{ bits/sec or bps} \\
150 \text{ bps} \\
300 \text{ bps} \\
600 \text{ bps} \\
1200 \text{ bps} \\
2400 \text{ bps} \\
4800 \text{ bps} \\
9600 \text{ bps} \\
19200 \text{ bps} \\
38400 \text{ bps}
\end{array}$$

The term *baud* is also used as a synonym for bits-per-second, but this is not proper. We will also assume that the terminal will be in the asynchronous mode of operation rather than the synchronous mode. In fact, most terminals do not support the synchronous mode of communications.

The Pro-Log 7304 Dual UART card (Fig. 6-13) contains two USARTs, which means that the board can be used to communicate with two terminals, or even two other computers, at the same time, independently. Both channels (A and B) can use the RS-232C protocol and, in addition, channel B can use the 20-mA protocol. The board also contains a programmable bit-rate generator. Thus, the bits and bit rates given in Table 6-12 can be used with the USARTs. This interface can also be used in DTE (Data Terminal Equipment) and DCE (Data Communications Equipment) applications. Usually, computers and modems are two types of DCE and most of the peripherals, such as terminals and printers, are DTE. Thus, the 7304 card would normally be configured as a DCE and your terminal would be a DTE.

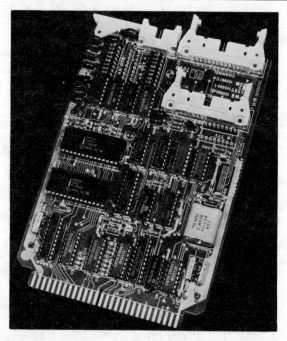

FIGURE 6-13
The Pro-Log 7304 dual UART card. *(Courtesy Pro-Log Corp.)*

You must not connect two DCE or DTE devices together, however, or you will burn out some of the chips in one or the other of the devices.

Before we can use this board with any type of terminal, the device addresses that the board uses will have to be established, and some software will have to be written. Because of the way that it was designed, each USART contains two input and two output ports. One of the input ports is used to input receiver data and one of the output ports is used to output data to the transmitter. The other two ports are used to control the operation of the USART. The card also contains a bit-rate generator, which looks like an output port. By writing an 8-bit value to this port, the bit rates used by the A and B channels can be controlled independently. This port is also used to enable and disable interrupts generated by the USARTs (Fig. 6-14). Thus, the 7304 card contains 5 output ports and 4 input ports. There is no input port with the same address as the bit-rate generator output port. Because of the way in which the board was designed, 8 device addresses are used by this card.

Table 6-12. Bit Rates Available on the Pro-Log 7304 Dual UART Card

Bit Rate	Control Bits
9600 bps	111
4800 bps	110
2400 bps	101
1800 bps	100
1200 bps	011
300 bps	010
150 bps	001
110 bps	000

When the 7304 card is shipped from the factory, the SX and SY address jumpers are positioned such that the first address used by the board is E0. Since the board uses up 8 addresses, the last address used is E7. The SX jumper is used to select 1 of 4 possible states for A7 and A6, and the SY jumper is used to select 1 of 8 possible states for A5, A4, and A3. Since the card uses 8 device addresses, A2, A1, and A0 are decoded on the board to either select one of the USARTs or the bit-rate generator logic. The various jumper positions and the device addresses used are summarized in Table 6-13. Since the 7304 card comes from the factory jumpered for addresses E0–E7, which jumpers are in place? The jumpers SX-3 and SY-4 are soldered on the board. Incidentally, the board can only be used as an accumulator I/O device, and the IOEXP* signal must be a logic 0.

The relationship between the port addresses and the USARTs and the bit-rate generator is summarized in Table 6-14. As you can see, the USARTs have data and control ports, and the bit-rate generator looks like the simple output port that it is (an SN74LS273 8-bit latch).

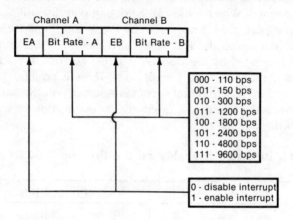

FIGURE 6-14
The structure of the bit-rate generator output port.

Unlike any of the previous interfaces discussed, USARTs have to be *programmed* before they can be used to communicate with a terminal. This means that values have to be output to the USART that will configure the USART for different modes of operation. These values *are not* transmitted to the terminal or receiving device.

Table 6-13. Jumper/Address Assignments for the 7304 Dual UART Card

		SY							
		0	1	2	3	4	5	6	7
	0	00-07	08-0F	10-17	18-1F	20-27	28-2F	30-37	38-3F
SX	1	40-47	48-4F	50-57	58-5F	60-67	68-6F	70-77	78-7F
	2	80-87	88-8F	90-97	98-9F	A0-A7	A8-AF	B0-B7	B8-BF
	3	C0-C7	C8-CF	D0-D7	D8-DF	E0-E7	E8-EF	F0-F7	F8-FF

Table 6-14. Function/Port Assignments on the Pro-Log 7304 Card

Port Address	Function	
	Read	Write
XXXXX111	Not used	Not used
XXXXX110	Not used	Not used
XXXXX101	Not used	Not used
XXXXX100	Status of channel A	Mode/command of channel A
XXXXX011	Receiver data of channel A	Transmitter data of channel A
XXXXX010	Status of channel B	Mode/command of channel B
XXXXX001	Receiver data of channel B	Transmitter data of channel B
XXXXX000	Not used	Bit rate/interrupt register

In fact, two values have to be output to the USART and to the *same* output port address. This may seem strange, but the USART has been designed to accept this information and use it internally. Of course, the two values have to be output in a specific order. The first value that the USART expects is the Mode word. Its structure is shown in Fig. 6-15. The Mode word is used to program the USART for the number of stop bits that will be transmitted, and which it will expect to receive, after the MSB of the data value has been either transmitted or received. The two parity bits let us use either even or odd parity, if the parity logic is enabled. This gives us the ability to determine if a bit has been lost in the communications link or if a bit has changed state. We can also program the USART to operate on 5-, 6-, 7-, or 8-bit values. Ordinarily, since we are using an 8-bit microprocessor, the USART would be programmed to transmit and receive 8-bit values. The USART's baud-rate factor is selectively based on the frequencies generated by the bit-rate generator on the card. As an example, a Mode word of 11001110 (CE) might be used. This programs the USART for 2 stop bits, no parity bit, 8-bit characters, and a 16× baud rate.

Once the Mode word has been output, the Command word has to be output to the same port. The structure of this word can be seen in Fig. 6-16. In one application, a value of 00010101 (15) might be

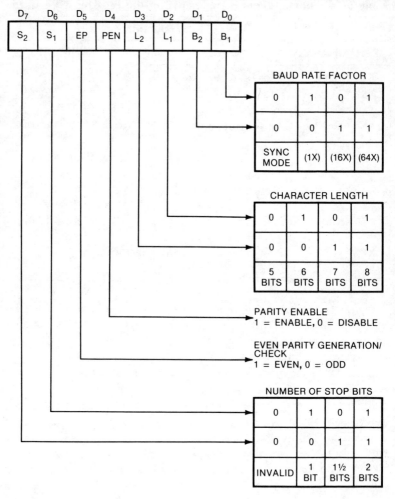

FIGURE 6-15

Mode word structure for the 8251A USART. *(Courtesy Intel Corp.)*

output as the Command word. This resets the error flags and enables both the receiver and the transmitter in the USART.

Once the USART has been programmed and a bit-rate selection word has been output, the USART can be used to receive and transmit data. How will we know when a character has been received by the USART? The USART has a status word, where one bit represents

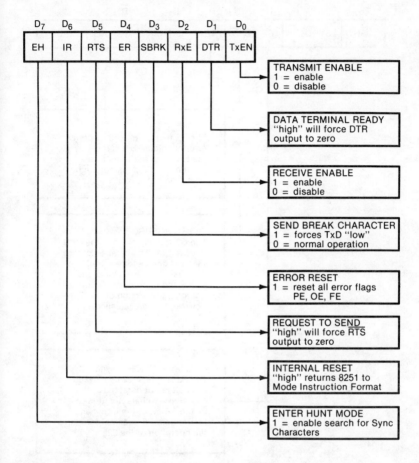

FIGURE 6-16

Structure of the Command word for the 8251A USART. *(Courtesy Intel Corp.)*

whether or not the receiver has received a character. Another bit in this same word tells us whether the transmitter is busy or not busy. Once the transmitter is in the "not busy" state, another character can be output to it. The position of these two flags in the 8-bit status word is shown in Fig. 6-17. The transmitter-ready flag (TxRDY) is a logic 0 when the transmitter is busy, and a logic 1 when it is not busy. The receiver-ready flag (RxRDY) is a logic 0 if no character has been received, and a logic 1 when a character has been received.

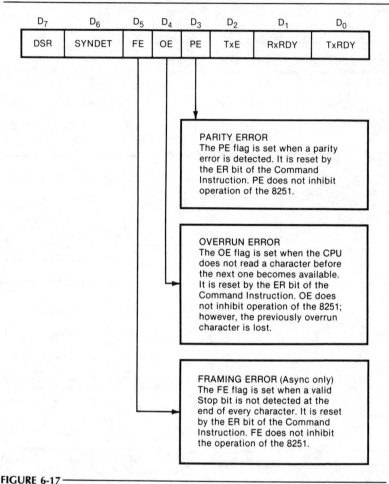

FIGURE 6-17

The Status (Flag) word of the 8251A USART. *(Courtesy Intel Corp.)*

The status word also contains 3 error flags; the parity, the overrun, and the framing error flags. Note that these flags are all reset, when the status word is read into the microprocessor.

With all of this hardware behind us, the software is almost trivial. Examples 6-17 and 6-18 program both the A and B channels for 2400-bps operation and no interrupts, and also program the channel A USART for 2 stop bits, no parity bit, 8 data bits, and a 16× baud rate. The receiver and transmitter are both enabled (Command word)

Example 6-17. Testing the Crt Screen.

1. Using the 8085 microprocessor:

TEST:	MVI A,55H	;01010101
	OUT E0H	;2400 bps
	MVI A,CEH	;11001110
	OUT E4H	;Channel A mode
	MVI A,15H	;00010101
	OUT E4H	;Channel A command
NXTLN:	MVI B,A0H	;ASCII space
LOOP	CALL CRTOUT	;Display register B
	INR B	;Character + 1
	MOV A,B	;Get character in A
	CPI E0H	;Done all 64?
	JNZ LOOP	;No, do another
	MVI B,0DH	;Carriage return
	CALL CRTOUT	;Display it
	MVI B,0AH	;Line feed
	CALL CRTOUT	;Display it
	JMP NXTLN	;Do another line
CRTOUT	IN E4H	;Get flags
	ANI 01H	;Save D0
	JZ CRTOUT	;Flag is 0, loop
	MOV A,B	;Get character
	OUT E3H	;Transmit it
	RET	

2. Using the Z-80 microprocessor:

TEST	LD A,55H	;01010101
	OUT (E0H),A	;2400 bps
	LD A,CEH	;11001110
	OUT (E0H),A	;Channel A mode
	LD A,15H	;00010101
	OUT (E4H),A	;Channel A command
NXTLN	LD B,A0H	;ASCII space
LOOP	CALL CRTOUT	;Display register B
	INC B	;Character + 1

Example 6-17. Cont.

```
              LD A,B           ;Get character in A
              CP E0H           ;Done all 64?
              JR NZ,LOOP       ;No, do another
              LD B,0DH         ;Carriage return
              CALL CRTOUT      ;Display it
              LD B,0AH         ;Line feed
              CALL CRTOUT      ;Display it
              JP NXTLN         ;Do another line

    CRTOUT    IN A,(E4H)       ;Get flags
              AND A,01H        ;Save D0
              JR Z,CRTOUT      ;Flag is 0, loop
              LD A,B           ;Get character
              OUT (E3H),A      ;Transmit it
              RET
```

and the error flags are reset. Once this is done, the microprocessor outputs all of the ASCII characters with values between A0 and DF, and follows this with a carriage return and a line feed. This simple program can be used to make sure that your terminal receives characters properly.

Examples 6-19 and 6-20 contain some general-purpose terminal subroutines that you can use in your own software. We have assumed in these examples that both the baud rate and the USART have already been programmed.

There are four general-purpose subroutines that you can use in your own software. The CRTEST (crt test) subroutine simply tests the keyboard flag and returns with the result reflected in the A register and flags. This subroutine might be called periodically to make sure that, in long programs, keyboard characters aren't missed. You might save the A register at the beginning of the subroutine and restore it at the end so that only the flags are altered. The CRTNE (crt no echo) subroutine will wait in a loop until a key is pressed and, then, the microprocessor will return with the character in the A register or accumulator. The AND instruction clears the parity bit (D7) in the character to zero. The keyboard character is not transmitted to the crt, line printer, etc. The CRTGP (crt general-purpose) subroutine will wait for

Example 6-18. Testing the Crt Screen.

1. Using the 6502 microprocessor:

TEST	LDA #$55	;01010101
	STA $C0E0	;2400 bps
	LDA #$CE	;11001110
	STA $C0E4	;Channel A mode
	LDA #$15	;00010101
	STA $C0E4	;Channel A command
NXTLN	LDX #$A0	;ASCII space
LOOP	JSR CRTOUT	;Display register
	INX	;Character + 1
	CPX #$E0	;Done all 64?
	BNE LOOP	;No, do another
	LDX #$0D	;Yes, carriage return
	JSR CRTOUT	;Display it
	LDX #$0A	;Line feed
	JSR CRTOUT	;Display it
	JMP NXTLN	;Do another line
CRTOUT	LDA $C0E4	;Get the flags
	AND #$01	;Save D0
	BEQ CRTOUT	;Flag is 0, loop
	STX $C0E3	;Output character
	RTS	

2. Using the 6800 microprocessor:

TEST	LDA A #$55	;01010101
	STA A $C0E0	;2400 bps
	LDA A #$CE	;11001110
	STA A $C0E4	;Channel A mode
	LDA A #$15	;00010101
	STA A $C0E5	;Channel A command
NXTLN	LDA B #$A0	;ASCII space
LOOP	JSR CRTOUT	;Display register
	INC B	;Character + 1
	CMP B #$E0	;Done all 64?
	BNE LOOP	;No, do another
	LDA B #$0D	;Yes, carriage return

Example 6-18. Cont.

```
            JSR CRTOUT      ;Display it
            LDA B #$0A      ;Line feed
            JSR CRTOUT      ;Display it
            BRA NXTLN       ;Do another line

  CRTOUT    LDA A $C0E4     ;Get the flags
            AND A #$01      ;Save D0
            BEQ CRTOUT      ;Flag is 0, loop
            STA B $C0E3     ;Output character
            RTS
```

a key to be pressed, get the character, and transmit it back to the crt. Finally, the CRTOUT subroutine can be called with the character to be displayed (transmitted) in the A register or accumulator.

If you want, you can use interrupts with the 7304 interface. To do this, the interrupts have to be enabled on the board (along with enabling the microprocessor's interrupt), by using bit D7 of the bit-rate output port to enable channel A interrupts, and bit D3 to enable channel B interrupts. If one of these bits is a logic 1, then, either the receiver or transmitter flag for the USART can generate an interrupt. The interrupt logic uses the INTRQ* line and a priority interrupt controller can be used with the 7304 (these flag signals are brought up to one of the user connectors). If a priority interrupt controller is not used, then, in your interrupt-service subroutine, you will have to examine the receiver and transmitter flags in both status words (if both USARTs are being used) to determine the source of the interrupt.

The USART chips used on the 7304 interface card are the 8251A, which is made by Intel Corporation, National Semiconductor, and AMD, among others. If you need more information about the 8251A USART chip, refer to one of the manufacturer's data sheets. Even though this chip was designed for the 8080 microprocessor, it will work in STD bus microcomputers that use the 6800, 6809, 6502, 8088, Z-80, and NSC800 microprocessors.

Example 6-19. General-Purpose Crt Subroutines.

1. Using the 8085 microprocessor:

CRTEST:	IN E4H	;Get flags
	ANI 02H	;Key pressed?
	RET	
CRTNE:	CALL CRTEST	;Get kybd flag
	JZ CRTNE	;Wait for a key
	IN E3H	;Get key value
	ANI 7FH	;Parity = 0
	RET	
CRTGP:	CALL CRTNE	;Get key value
CRTOUT	PUSH PSW	;Save character
CRT1:	IN E4H	;Get flags
	ANI 01H	;Transmitter busy?
	JZ CRT 1	;Yes, wait
	POP PSW	;Get character
	OUT E3H	;Output it
	RET	

2. Using the Z-80 microprocessor:

CRTEST:	IN A,(E4H)	;Get flags
	AND A,02H	;Key pressed?
	RET	
CRTNE:	CALL CRTEST	;Get kybd flag
	JR Z,CRTNE	;Wait for a key
	IN A,(E3H)	;Get key value
	AND A,7FH	;Parity = 0
	RET	
CRTGP	CALL CRTNE	;Get key value
CRTOUT	PUSH AF	;Save character
CRT1	IN A,(E4H)	;Get flags
	AND A,01H	;Transmitter busy?
	JR Z,CRT1	;Yes, wait
	POP AF	;Get character
	OUT (E3H),A	;Output it
	RET	

Example 6-20. General-Purpose Crt Subroutines.

1. Using the 6502 microprocessor:

CRTEST	LDA $C0E4	;Get flags
	AND #$02	;Save D1
	RTS	
CRTNE	JSR CRTEST	;Get kybd flag
	BEQ CRTNE	;Wait for a key
	LDA $C0E3	;Get the key value
	AND #$7F	;Parity = 0
	RTS	
CRTGP	JSR CRTNE	;Get a key value
CRTOUT	PHA	;Save the character
CRT1	LDA $C0E4	;Get flags
	AND #$01	;Transmitter busy?
	BEQ CRT1	;Yes, wait
	PLA	;Get character
	STA $C0E3	;Output it
	RTS	

2. Using the 6800 microprocessor:

CRTEST	LDA A $C0E4	;Get flags
	AND A #$02	;Save D1
	RTS	
CRTNE	JSR CRTEST	;Get kybd flag
	BEQ CRTNE	;Wait for a key
	LDA A $C0E3	;Get the key value
	AND #$7F	;Parity = 0
	RTS	
CRTGP	JSR CRTNE	;Get a key value
CRTOUT	PSH A	;Save the character
CRT1	LDA A $C0E4	;Get flags
	AND A #$01	;Transmitter busy?
	BEQ CRT1	;Yes, wait
	PUL A	;Get character
	STA A $C0E3	;Output it
	RTS	

The STD BUS Standard
Appendix A

The following STD BUS Specifications, with its tables, drawings, and other pertinent information has been furnished by courtesy of the Matrix Corporation.

INTRODUCTION

The STD BUS standardizes the physical and electrical aspects of modular 8-bit microprocessor card systems. It provides a dedicated and orderly interconnection scheme (Fig. A-1). The standardized pinout and 56-pin connector lend themselves to a bused motherboard that permits any card to work in any slot.

The STD BUS is dedicated to internal communications. All other interconnections are made via suitable connectors at the I/O interface card edge. The concept gives an orderly signal flow across the cards. Peripheral and I/O devices can be connected to the system, according to their own unique connector and cabling requirements.

ORGANIZATION AND FUNCTIONAL SPECIFICATIONS (WITH PIN DEFINITIONS)

The STD BUS pinout is organized into four functional groups:

- Dual Power Buses: Pins 1–6 and 53–56
- Data Bus: Pins 7–14
- Address Bus: Pins 15–30
- Control Bus: Pins 31–52

The organization and pinouts are shown in Table A-1. This table lists the mnemonic-function and signal-flow direction (referenced to the processor

Table A-1. STD BUS Pinouts With Signal Flow Referenced to the Processor Card

		Component Side				Circuit Side	
	Pin	Mnemonic	Signal Flow	Description	Pin	Mnemonic	Signal Flow	Description
LOGIC POWER BUS	1	+5 V dc	In	Logic Power (bused)	2	+5 V dc	In	Logic Power (bused)
	3	GND	In	Logic Ground (bused)	4	GND	In	Logic Ground (bused)
	5	VBB #1	In	Logic Bias No. 1 (−5 V)	6	VBB #2	In	Logic Bias No. 2 (−5 V)
DATA BUS	7	D3	In/Out	Low-Order Data Bus	8	D7	In/Out	High-Order Data Bus
	9	D2	In/Out	Low-Order Data Bus	10	D6	In/Out	High-Order Data Bus
	11	D1	In/Out	Low-Order Data Bus	12	D5	In/Out	High-Order Data Bus
	13	D0	In/Out	Low-Order Data Bus	14	D4	In/Out	High-Order Data Bus
ADDRESS BUS	15	A7	Out	Low-Order Address Bus	16	A15	Out	High-Order Address Bus
	17	A6	Out	Low-Order Address Bus	18	A14	Out	High-Order Address Bus
	19	A5	Out	Low-Order Address Bus	20	A13	Out	High-Order Address Bus
	21	A4	Out	Low-Order Address Bus	22	A12	Out	High-Order Address Bus
	23	A3	Out	Low-Order Address Bus	24	A11	Out	High-Order Address Bus
	25	A2	Out	Low-Order Address Bus	26	A10	Out	High-Order Address Bus
	27	A1	Out	Low-Order Address Bus	28	A9	Out	High-Order Address Bus
	29	A0	Out	Low-Order Address Bus	30	A8	Out	High-Order Address Bus

Table A-1. Cont.

		Component Side			Circuit Side	
	Pin	Mnemonic	Signal Flow	Description	Pin	Mnemonic	Signal Flow	Description
---	---	---	---	---	---	---	---	---
CONTROL BUS	31	WR*	Out	Write to Memory or I/O	32	RD*	Out	Read Memory or I/O
	33	IORQ*	Out	I/O Address Select	34	MEMRQ*	Out	Memory Address Select
	35	IOEXP	In/Out	I/O Expansion	36	MEMEX	In/Out	Memory Expansion
	37	REFRESH*	Out	Refresh Timing	38	MCSYNC*	Out	CPU Machine Cycle Sync
	39	STATUS 1*	Out	CPU Status	40	STATUS 0*	Out	CPU Status
	41	BUSAK*	Out	Bus Acknowledge	42	BUSRQ*	In	Bus Request
	43	INTAK*	Out	Interrupt Acknowledge	44	INTRQ*	In	Interrupt Request
	45	WAITRQ*	In	Wait Request	46	NMIRQ*	In	Nonmaskable Interrupt
	47	SYSRESET*	Out	System Reset	48	PBRESET*	In	Push-Button Reset
	49	CLOCK*	Out	Clock from Processor	50	CNTRL*	In	AUX Timing
	51	PCO	Out	Priority Chain Out	52	PCI	In	Priority Chain In
AUXILIARY POWER BUS	53	AUX GND	In	AUX Ground (bused)	54	AUXGND	In	AUX Ground (bused)
	55	AUX +V	In	AUX Positive (+12 V dc)	56	AUX −V	In	AUX Negative (−12 V dc)

*Low-level active indicator

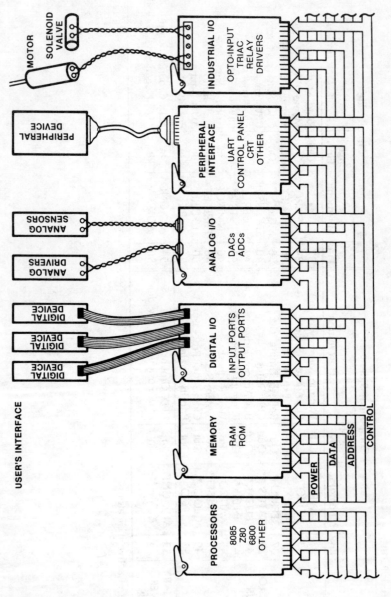

FIGURE A-1

STD BUS implementation.

card in control of the BUS) for each pin of the STD BUS. The STD BUS is further defined as requiring a 56-pin (dual 28) card-edge connector, with 0.125-inch pin centers. Connectors are on a spacing interval of 0.5-inch centers minimum, and they accept the standard 4.5 × 6.5 × 0.062-inch (11.43 cm × 16.51 cm × 0.16 cm) card.

Dual Power Buses (Pins 1-6 and 53-56)

The dual power buses accommodate logic and analog power distribution. As many as five separate power supplies can be used with two separate ground returns as shown in Table A-2.

Table A-2. Bus Power Distribution

Pin	Description	Comments
1 & 2	Logic power	Logic power source (+5 V dc)
3 & 4	Logic ground	Logic power return bus
5	Logic bias voltage	Low-current logic supply No. 1 (−5 V)
6	Logic bias voltage	Low-current logic supply No. 2 (−5 V)
53 & 54	Auxiliary ground	Auxiliary power return bus
55	Auxiliary positive	Positive dc supply (+12 V)
56	Auxiliary negative	Negative dc supply (−12 V)

Data Bus (Pins 7-14)

The data bus is an 8-bit, bidirectional, 3-state bus. (Bidirectional means signals may flow either into or out of any card on the bus.) Direction of data is normally controlled by the processor card via the control bus. The data direction is normally affected by such signals as read (RD*), write (WR*), and interrupt acknowledge (INTAK*).

The data bus uses high-level active logic. All cards are required to release the bus to a high-impedance state when not in use. The processor card releases the data bus in response to bus request (BUSRQ*) input from an alternate system controller, as in DMA transfers.

Address Bus (Pins 15-30)

The address bus is a 16-bit, 3-state, high-level active bus. Normally, the address originates at the processor card. The card releases the address bus in response to a BUSRQ* input from an alternate controller.

The address bus provides 16 address lines for decoding by either memory or I/O. Memory request (MEMRQ*) and I/O request (IORQ*) control lines distinguish between the two operations. The particular microprocessor that you use determines the number of address lines and how they are applied. This is shown in Table A-3.

Table A-3. Microprocessor Address Lines

Processor	No. of Mem Address Lines	Address Lines During Refresh	No. of I/O Address Lines	
			I/O-Mapped I/O	Memory-Mapped I/O
8080	16	—	Lower 8	16
8085	16	—	Lower 8	16
Z-80	16	Lower 7	Lower 8	16
6800	16	—	—	16
6809	16	—	—	16
6502	16	—	—	16
NSC800	16	Lower 7	Lower 8	16

Control Bus (Pins 31-52)

The control bus determines the flexibility of the STD BUS. Signal lines are grouped into five separate areas: memory and I/O control, peripheral timing, clock and reset, interrupt and bus control, and serial priority chain.

Memory and I/O Control—These lines provide the signals for fundamental memory and I/O operations. Simple applications may only require the following six control signals:

- WR*—Write to memory or I/O (3-state, active-low), pin 31.
  This signal indicates that the bus holds valid data to be written in the addressed memory or output device. WR* is the clock pulse, which writes data to memory or output port latches. The signal originates from the processor, which also provides the output data on the bus.
- RD*—Read from memory or I/O (3-state, active-low), pin 32.
  This signal indicates that the processor or other bus-controlling device needs to read data from memory or from an I/O device. The selected I/O device or memory utilizes this signal to gate data onto the bus. RD* originates from the processor, which accepts the data from the bus.
- IORQ*—I/O address select (3-state, active-low), pin 33.
  This signal indicates that the address lines hold a valid I/O address for an I/O read or write. It is used on the I/O cards and is gated with either RD* or WR* to designate I/O operations.
- MEMRQ*—Memory address select (3-state, active-low), pin 34.
  This signal indicates that the address bus holds a valid address for memory read or memory write operations. It is used on memory cards and is gated with either RD* or WR* to designate memory operations.
- IOEXP—I/O expansion (high expand, low enable), pin 35.
  This signal expands or enables I/O port addressing. An active-low enables primary I/O operations. An example of its use is to allow common address decoding in memory-mapped I/O operations. Simple systems can generally strap this signal to ground.
- MEMEX—Memory expansion (high expand, low enable), pin 36.
  This signal expands or enables memory addressing. An active-low en-

ables the primary system memory. MEMEX allows memory overlay such as that found in bootstrap operations. A control card may switch out the primary system memory to make use of an alternate memory. Simple systems can generally strap this signal to ground.

Peripheral Timing Control—These lines provide control signals that enable the use of the STD BUS with microprocessors that service their own peripheral devices. The STD BUS is intended to service any 8-bit microprocessor. Most peripheral devices work only with the microprocessor they are designed for. Four control lines of the STD BUS are designated for peripheral timing. They are defined specifically for each type of microprocessor, so that it can best serve its own peripheral devices. In this way, the STD BUS is not limited to only one processor.

- REFRESH*—(3-state, active-low), pin 37.
  This signal refreshes dynamic memory. It may be generated on the processor card or on a separate control card. The nature and timing of the signal may be a function of the memory device or of the microprocessor. In systems without refresh, this signal can be any specialized memory control signal. Simple systems with static memory may disregard REFRESH*.
- MCSYNC*—Machine cycle sync (3-state, active-low), pin 38.
  This signal occurs once during each machine cycle of the processor. (Machine cycle is defined as the sequence that involves addressing, data transfer, and execution.) MCSYNC* defines the beginning of the machine cycle. The exact nature and timing of this signal are processor-dependent. MCSYNC* keeps specialized peripheral devices synchronized with the processor's operation. It can also be used for controlling a bus analyzer, which can analyze bus operations cycle by cycle.
- STATUS 1*—Status control line 1 (3-state, active-low), pin 39.
  This signal provides secondary timing for peripheral devices. When available, STATUS 1* is considered as a signal for identifying instruction fetch.
- STATUS 0*—Status control line 0 (3-state, active-low), pin 40.
  This signal provides additional timing for peripheral devices. Table A-4 defines the peripheral timing-control lines for various 8-bit microprocessors. The designated pins are also being defined for other microprocessors.

Interrupt and Bus Control—These lines allow the implementation of such bus control schemes as direct-memory access, multiprocessing, single stepping, slow memory, power-fail-restart, and a variety of interrupt methods. The STD BUS includes provision for a serial priority chain. Parallel priority schemes can also be implemented.

- BUSAK*—Bus acknowledge (active-low), pin 41.
  This signal indicates that the bus is available for use by a requesting

Table A-4. Peripheral Timing-Control Lines for Various 8-Bit Microprocessors

Processor	REFRESH* Pin 37	MCSYNC* Pin 38	STATUS 1* Pin 39	STATUS 0* Pin 40
8080	—	SYNC*	M1*	—
8085	—	ALE*	S1*	SO*
NSC800	REFRESH*	ALE*	S1*	SO*
8088	—	ALE*	DT/R*	SSO*
Z-80	REFRESH*	(RD*+WR* +INTAK*)	M1*	—
6800	—	Ø2*	VMA*	R/W*
6809	—	EOUT* (Ø2*)	—	R/W*
6809E	—	EOUT* (Ø2*)	LIC*	R/W*
6502	—	Ø2*	SYNC*	R/W*

```
*      Low-level active          R/W*   Read high, write low
—      Not used                  DT/R*  Data transmit high, receive low
```

controller. The controlling processor responds to a BUSRQ* by releasing the bus and giving an acknowledge signal on the BUSAK* line. BUSAK* occurs at the completion of the current machine cycle.

- BUSRQ*—Bus request (active-low, open collector), pin 42.

  This signal causes the controller processor to suspend operations on the STD BUS by releasing all 3-state STD BUS lines for use by another processor. The STD BUS is released when the current machine cycle has been completed. BUSRQ* is used in applications requiring direct-memory access (DMA). In complex systems, it can be an input, or an output, or it can be bidirectional, depending on the supporting hardware.

- INTAK*—Interrupt acknowledge (active-low), pin 43.

  This signal tells the interrupting device that the processor card is ready to respond to the interrupt. For vectored interrupts, the interrupting device places the vector address on the data bus during INTAK*. This signal can be combined with a priority signal, if multiple controllers need bus access. INTAK* is not used in nonvectored interrupt schemes.

- INTRQ*—Interrupt request (active-low, open collector), pin 44.

  This processor-card input signal conditionally interrupts the program. It is masked and ignored by the processor, unless deliberately enabled by a program instruction. If the processor accepts the interrupt, it usually acknowledges by dropping INTAK* (pin 43). Other actions depend on the specific type of processor, the interrupt-related program instructions, and the hardware support of the interrupt mechanism.

- WAITRQ*—Wait request (active-low, open collector), pin 45.

  This input signal to the processor suspends operations as long as it remains low. Normally, the processor holds in a state that maintains a valid address on the address bus. WAITRQ* can be used to insert "wait" states in the processor cycle. Examples of its use include slow-memory operations and single stepping.

- NMIRQ*—Nonmaskable interrupt (active-low, open collector), pin 46.
  This signal is a processor-card interrupt input of the highest priority. It should be used for critical processor signalling, e.g., power-fail indications.

Clock and Reset—These lines provide the STD BUS with basic clock timing and reset capability.
- SYSRESET*—System reset (active-low), pin 47.
  This signal is an output from the system reset circuit, which is triggered by power-on detection, or by the push-button reset. The system reset bus line should be applied to all bus cards that have latch circuits requiring initialization.
- PBRESET*—Push-button reset (active-low), pin 48.
  This signal is an input line to the system reset circuit.
- CLOCK*—Clock from processor, pin 49.
  This signal is a buffered processor clock signal, for use in system synchronization or as a general clock source.
- CNTRL*—Control, pin 50.
  This signal is an auxiliary circuit for special clock timing. It may be a multiple of the processor clock signal, a real-time clock signal, or an external input to the processor.

Serial Priority Chain—These lines are provided for interrupt or bus control. Two bus pins are allocated to the chain, which requires logic on the card to implement the priority function. Cards not needing the chain must jumper PCI to PCO on the card, if they are to be used in a serial priority scheme.
- PCO—Priority chain out, pin 51.
  This signal is sent to the PCI input of the next lower card in priority. A card that needs priority should hold PCO low.
- PCI—Priority chain in, pin 52.
  This signal is provided directly from the PCO of the next higher card in priority. A high level on PCI gives priority to the card sensing the PCI input.

ELECTRICAL SPECIFICATIONS

Absolute Maximum Ratings

The maximum ratings for the STD BUS card edge-connector pins, which are listed in Table A-5, are not recommended operating conditions. Above these values, damage to card components is possible. The specific voltage at which damage occurs is component-dependent.

It should be noted that unless otherwise specified, the removal of circuit cards that are compatible with the STD BUS, or the removal of their component parts from sockets, is not recommended while operating voltages are applied.

Table A-5. Pin Ratings

Parameter	Limit	Reference
Positive voltage applied to logic input or disabled 3-state output.	+5.5 V	GND pins 3, 4
Negative dc voltage applied to a logic input or disabled 3-state output.	−0.4 V	

Power Bus Voltage Tolerances

STD BUS cards normally require +5 V for logic operations. However, other operating voltages may be needed, according to individual card function and device types. Table A-6 shows the STD BUS power buses and voltage values. Note that these voltage values are specified at the card pins, not at the backplane traces.

Table A-6. STD BUS Power Buses and Voltage Values

Card Pin	Supply Voltage	Tolerance	Reference
1,2	VCC (+5 V)	±0.25 V	GND pins 3, 4
5	VBB #1 (−5 V)	±0.25 V	GND pins 3, 4
6	VBB #2 (−5 V)	±0.25 V	GND pins 3, 4
55	AUX +V (+12 V)	±0.5 V	AUX GND pins 53, 54
56	AUX −V (−12 V)	±0.5 V	AUX GND pins 53, 54

Logic Signal Characteristics

The STD BUS is designed for compatibility with industry-standard TTL logic. The specifications in Table A-7 apply over the specified temperature range for the STD BUS.

Table A-7. Logic Signal Characteristics

STD Bus Card Parameter	Test Conditions	Min	Max	Units
VOH (high-state output voltage)	VCC=MIN, IOH=−15 mA	2.4	—	V
VOL (low-state output voltage)	VCC=MIN, IOL=24 mA	—	0.5	V
VIH (high-state input voltage)		2.0	—	V
VIL (low-state input voltage)		—	0.8	V
tR, tF (rise time, fall time)		4	100	NS

MECHANICAL SPECIFICATIONS

The circuit card size and outline of the STD BUS are defined in Table A-8 and Figs. A-2 through A-5. The dimensions exclude the card ejector and I/O interface connections.

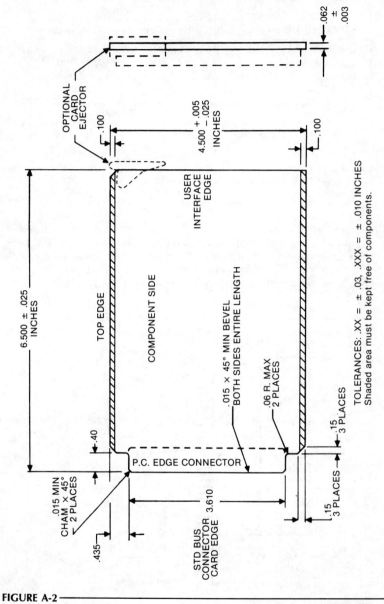

FIGURE A-2

STD BUS card outline (decimal).

Table A-8. STD BUS Card Dimensions

STD Card Dimensions	Inches	
	Nominal	Tolerance
Card Length	6.500	±0.025
Card Height	4.500	+0.005, −0.025
Plated Board Thickness	0.062	±0.003
Card Spacing	0.500	MIN

STD Card Dimensions	Millimeters	
	Nominal	Tolerance
Card Length	165.10	±0.64
Card Height	114.30	+0.13, −0.64
Plated Board Thickness	1.58	±0.08
Card Spacing	12.70	MIN

Minimum card spacing requires a consideration for component height, lead protrusion, and card clearance, in addition to the board thickness. Table A-9 lists recommended dimensions for these parameters; however, trade-offs can be made between component height and lead protrusion. Cards not meeting these requirements may need multiple card slot positions.

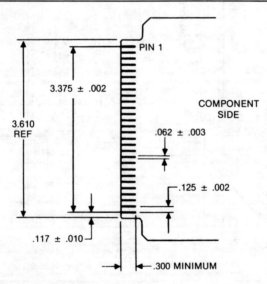

FIGURE A-3
STD BUS edge card finger design (decimal).

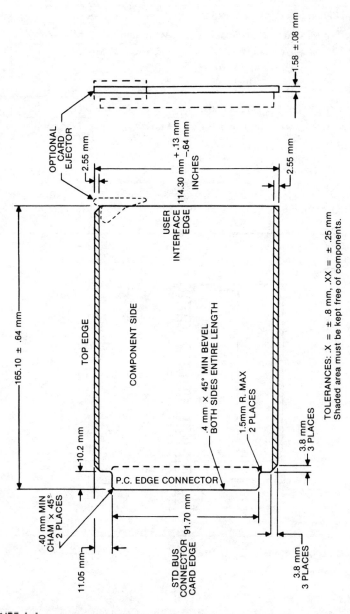

FIGURE A-4
STD BUS card outline (metric).

Table A-9. STD BUS Profile Dimensions for Minimum Spacing

Recommended Dimensions	Inches	
For Minimum Card Spacing	Maximum	Minimum
Component Height	0.375	—
Component Lead Protrusion[1]	0.040	—
Adjacent Card Clearance	—	0.010

Recommended Dimensions	Millimeters	
For Minimum Card Spacing	Maximum	Minimum
Component Height	9.52	—
Component Lead Protrusion[1]	1.02	—
Adjacent Card Clearance	—	0.25

[1]The card ejector occupies the top 1.4-inch (35.6 mm) of the card and protrudes 0.1-inch (2.54 mm) on each side of the card.

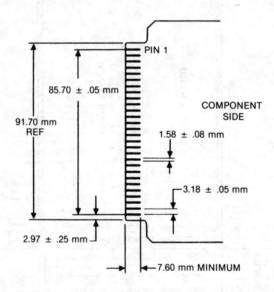

FIGURE A-5
STD BUS edge card finger design (metric).

Voltage Input Configurations
Appendix B

The following information on the RTI-1260 12-bit analog input board, along with the drawings, is furnished courtesy of Analog Devices.

INPUT MULTIPLEXER GUIDELINES

This RTI series product employs a very versatile data-acquisition subsystem that can be user-configured for maximum effectiveness in a wide variety of applications. The following sections describe the choices available and provide all the information necessary to permit the user to tailor the data-acquisition hardware to suit his own requirements. These sections are concerned primarily with analog inputs and the microcomputer interface.

ANALOG INPUT MULTIPLEXER

As shown in Fig. B-1, there are three basic ways to configure the input multiplexer. The choice will depend upon the application.

1. *Single-ended input* (Fig. B-1A) uses single-pole switching to select the desired input signal and, therefore, maximizes the available number of input channels. However, this configuration provides no rejection of common-mode noise and measures all inputs with respect to the local analog common. It is, therefore, best suited for use with relatively large signals (greater than 1-volt full scale) and in situations where the input signals originate close to the RTI card. Moreover, it is best used with low-impedance sources, since large source resistances will cause errors due to the instrumentation amplifier's input bias current.
2. *Pseudo-differential input* (Fig. B-1C) improves upon the single-ended scheme in one respect: it measures the input signals with respect to a

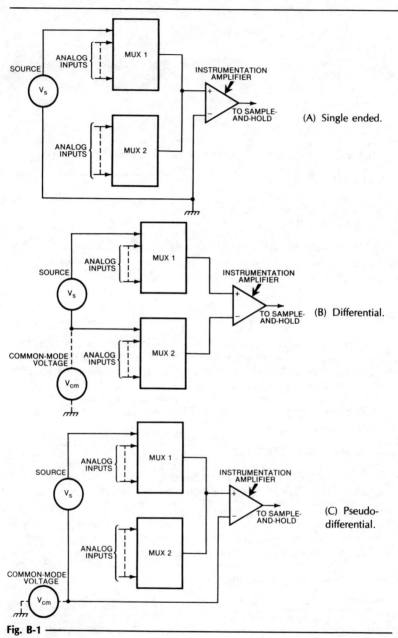

Fig. B-1

Input multiplexer configuration.

common point at their source. It is still necessary that all signals have a common terminal, but the effects of voltage differences between the local common and the signal common are eliminated, and noise pickup in signal cables will be reduced to some extent. Thus, the signal sources can be farther from the multiplexer than would be desirable with single-ended operation, and the full complement of input channels is still available. Bias current errors will still exist if high source resistances are used. This input structure is useful in applications where it is necessary to reduce the effects of interchassis offsets, or other errors common to all input signals.

3. *Full differential input* (Fig. B-1B) is most effective in reducing the effects of noise and bias current, but sacrifices channel capacity. As shown in Fig. B-1C, a double-pole arrangement is used to switch both the high and low side of each signal source. Thus, the input signals need not have a common point; each can "float" within the common-mode limits of the instrumentation amplifier. Common-mode noise voltages will be rejected by the instrumentation amplifier, and the symmetry of the input system will enhance ac noise rejection. If the source resistances seen at each lead of a signal pair are approximately equal, then any error due to bias currents will tend to cancel, leaving only input offset current as an error source. Thus, full differential operation should be considered for any application involving low-level input signals, high source resistances, or significant amounts of common-mode noise, such as might result from long signal cables.

In connection with full differential operation, it is important to remember that a return path must be provided for the input bias currents of the instrumentation amplifier. In cases where the signal source is floating (that is, it has no galvanic connection to the RTI card's analog common), a suitable path must be provided. Fig. B-2 shows how the signal-cable shield can be used for this purpose.

Although the use of full differential multiplexing can minimize the effects of source resistance, system accuracy will generally be much better if large source resistances can be avoided. A signal with an effective resistance of 100 ohms will be quite vulnerable to noise contamination on even a short cable run and, above 10,000 ohms, the settling time of the input multiplexer will be degraded. Selection of the input multiplexer configuration on the RTI card is accomplished by means of wire-wrap jumpers.

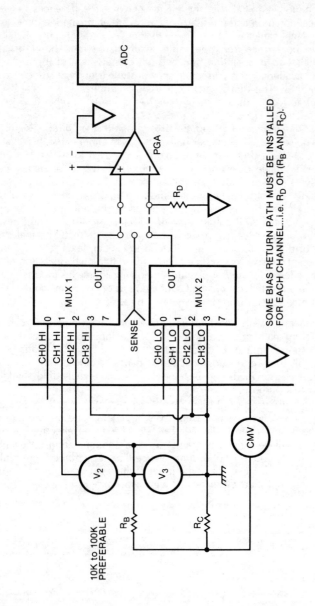

FIGURE B-2
Using signal-cable shield to provide a current return path.

Index Of STD BUS Manufacturers
APPENDIX C

AERCO
Box 18093
Austin, TX 78760-8093
(512) 385-7405

AMTEK
12740 28th N.E.
Seattle, WA 98125
(206) 363-0217

Analog Devices
Rt. 1, Industrial Park
Norwood, MA 02062
(617) 329-4700

Antona Corporation
13600 Ventura Blvd., Suite A
Sherman Oaks, CA 91423
(213) 986-6651

Applied Micro Technology, Inc.
P.O. Box 3042
Tuscon, AZ 85702
(602) 622-8605

Applied Systems Corp.
26401 Harper Ave.
St. Clair Shores, MI 48081
(313) 779-8700

Arcturus
P.O. Box 271
Wayland, MA 01778
(617) 358-4319

Atec, Inc.
P.O. Box 19426
Houston, TX 77024
(713) 468-7971

Augat
33 Perry Ave.
Attleboro, MA 02703
(617) 222-2202

Baradine Products, Ltd.
P.O. Box 86757
N. Vancouver, B.C.
Canada V7L 4L3
(604) 988-9853

Buckminster Corporation
99 Highland Ave.
Somerville, MA 02143
(617) 864-2456

Campbell Scientific, Inc.
P.O. Box 551
Logan, UT 84321
(801) 753-2342

Circuits and Systems, Inc.
2 Main St.
Hollis, NH 03049
(603) 465-7063

Computer Dynamics, Inc.
105 Main St.
Greer, SC 29651
(803) 877-7471

Contemporary Control Systems
935 Curtiss St.
Downers Grove, IL 60515
(312) 963-7070

Cytec Corp.
107 N. Washington St.
E. Rochester, NY 14506
(716) 391-4740

Data Translation
100 Locke Dr.
Marlboro, MA 01752
(617) 481-3700

Datricon Corporation
7911 NE 33rd Dr.
Portland, OR 97211
(503) 284-8277

Digital Dynamics
830 E. Evelyn
Sunnyvale, CA 94086
(408) 733-4660

Douglas Electronics
718 Marina Blvd.
San Leandro, CA 94577
(415) 483-8770

dy-4 Systems, Inc.
1573 Laperriere Ave.
Ottawa, K1Z 7T3
Canada
(613) 728-3711

E-L Instruments
61 First St.
Derby, CT 06418
(203) 735-8774

Electrologic, Inc.
1359 28th St.
Signal Hill, CA 90806
(213) 595-0551

Enlode Inc.
728 Kingsley Ave.
Orange Park, FL 32073
(904) 264-4405

Enterprise Systems Corporation
P.O. Box 643
Dover, NH 03820
(603) 742-7363

Euteknic Associates
1488 Ramon Dr.
Santa Clara, CA 95051
(408) 980-1096

Forethought Products
87070 Dukhobar Rd.
Eugene, OR 97402
(503) 485-8575

G and G Engineering
13708 Doolittle Dr.
San Leandro, CA 94577

Godbout Electronics
P.O. Box 2355
Oakland Airport, CA 94614
(415) 562-0636

Gordos Corporation
250 Glenwood Ave.
Bloomfield, NJ 07003
(201) 743-6800

GW3, Inc.
7239 Belinger Ct.
Springfield, VA 22150
(703) 451-2043

Hartmann-Lang
P.O. Box 693
Chatham, NJ 07928
(201) 635-7207

Intermagnetics General Corp.
P.O. Box 566
Guilderland, NY 12804
(518) 456-5456

Intersil, Inc.
10710 Tantau Ave.
Cupertino, CA 95014
(408) 996-5000

Intra Computer, Inc.
120-10 Audley St.
Kew Gardens, NY 11415
(212) 847-1936

I/O Controls, Inc.
517 E. Lincoln Hwy.
Langhorne, PA 19047
(215) 757-2772

Ironics, Inc.
117 Eastern Heights Dr.
Ithaca, NY 14850
(607) 277-4060

JF Microsystems
Star Route 1, Box 1174-D
Pasco, WA 99301
(509) 547-3397

Jonos, Ltd.
920-C E. Orangethorpe
Anaheim, CA 92801
(714) 871-1082

Kennedy Co.
1600 Shamrock Ave.
Monrovia, CA 91016
(213) 357-8831

Lang Systems
1392 Borregas Ave.
Sunnyvale, CA 94086
(408) 734-3332

LDI Pneutronics
2 Lomar Park
E. Pepperell, MA 01437
(617) 433-2721

Matrix Corporation
1639 Green St.
Raleigh, NC 27603
(919) 833-2837

Matrox Electronic Systems, Inc.
5800 Andover Ave.
T.M.R., Quebec, H4T 1H4
Canada
(514) 735-1182

Micro-Aide Corp.
482 W. Arrow H'way, Suite F
San Dimas, CA 91773
(213) 512-3804

Micro-Link Corporation
624B Range Line Rd.
Carmel, IN 46032
(317) 846-1721

Micro Source, Inc.
591 N. Clayton Rd.
New Lebanon, OH 45345
(513) 854-3266

Micro/Sys Inc.
P.O. Box 516
La Canada, CA 91011
(213) 790-7267

Microcomputer Systems, Inc.
1814 Ryder Dr.
Baton Rouge, LA 70808
(504) 769-2154

Micronet, Ltd.
P.O. Box 7066
Halifax, Nova Scotia
Canada B3K 5J4
(902) 429-3024

Miller Technology
16930 Sheldon Rd.
Los Gatos, CA 95030
(408) 395-2999

Mimic Electronics Co.
P.O. Box 921
Action, MA 01720
(617) 263-2101

Mostek Corporation
1215 W. Crosby Rd.
Carrollton, TX 75006
(214) 323-6000

Mullen Computer Products, Inc.
Box 6214
Hayward, CA 94545
(415) 783-2866

Northwest Microcomputer Systems
749 River Ave.
Eugene, OR 97404
(503) 688-7100

Octagon Systems Corp.
5150 W. 80th Ave.
Westminster, CO 80030
(303) 426-8540

PC/M, Inc., Bubbl-Tec Div.
6800 Sierra Court
Dublin, CA 94566
(415) 829-8700

Pro-Log Corporation
2411 Garden Rd.
Monterey, CA 93940
(408) 372-4593

Quasitronics, Inc.
211 Vandale Dr.
Houston, PA 15342
(412) 745-2663

Robinson-Nugent, Inc.
800 E. 8th St.
New Albany, ID 47150

Robotrol Corp.
1250 Oakmead Pkwy., Suite 210
Sunnyvale, CA 94086
(408) 732-8813

Samco
4417 Longworth
Alexandria, VA 22309
(703) 360-7019

Sensoray Corp.
780 E. Trimble Rd., Bldg. 608
San Jose, CA 95131
(408) 262-7271

Sibthorp Systems, Inc.
8050 Production Ave.
Florence, KY 41052
(606) 525-1300

Solar Wind Systems, Inc.
94 Galli Dr.
Novato, CA 94947
(415) 883-0404

Spurrier Peripherals
10513 LaMarie Dr.
Cincinnati, OH 45241
(513) 563-2625

Standard National Corporation
212 Main St.
North Redding, MA 01864
(617) 942-0514

Systek
6515 W. Clearwater Ave.
Kennewick, WA 99336
(509) 735-1200

System Service
3627 Longview Valley Rd.
Sherman Oaks, CA 91423
(213) 986-6651

Tetronics
322 E. Deepdale Rd.
Phoenix, AZ 85022
(602) 866-1926

TL Industries, Inc.
2573 Tracy Rd.
Northwood, OH 43619
(419) 666-8144

Transwave Corp.
Rd 1, Box 489
Vanderbilt, PA 15486
(412) 628-6370

Vector Electronic
12460 Gladstone Ave.
Sylmar, CA 91342
(213) 365-9661

Vero Electronics, Inc.
171 Bridge Rd.
Hauppauge, NY 11788
(516) 234-0400

Ward Systems, Inc.
P.O. Box K
Sumner, WA 98390

Whedco, Inc.
6107 Jackson Rd.
Ann Arbor, MI 48103
(313) 665-5473

Wintech Systems
Box 121361
Arlington, TX 76012
(817) 274-7553

Xitex Corp./QC Micro Systems
9861 Chartwell Dr.
Dallas, TX 75243
(214) 349-2490

XYZ Electronics, Inc.
Rt. 12, Box 322
Indianapolis, IN 46236
(317) 335-2525

Ziatech Corporation
2410 Broad St.
San Luis Obispo, CA 93401
(805) 541-0488

Zydeco, Inc.
1058 North Blvd.
Baton Rouge, LA 70821
(504) 383-1719

Index

A

Accumulator I/O, 30-31, 32, 41-42, 43, 60, 61, 66, 76, 79, 120, 156
ADC, 214-216; see also analog-to-digital converter
  software, 222-223
Address(es)
  and control signals, 41-42
  bus, 263
  decoding
    circuitry, 148
    designs, 60
    using gates for, 42-50
  detection
    circuitry, 66, 74
    logic, 101-102
  memory, 11, 61, 63
Addressing
  I/O device, 41-80
  nonabsolute, 55, 58
AND gate, 43-45
Analog
  input multiplexer, 273-275
  signal, 141, 145
  -to-digital converter(s), 141, 214
    interface, 141-145
  voltage, 103-104
ASCII keyboard interface, 130-132, 136
ATEC 710 thumbwheel switch interface, 231-233

B

Base interrupt operation, 162-163
Bit rate(s), 245, 246
  generator, 246, 247, 248
Bus(es), 263-264
  control, 265-266

Bus(es) — cont
  8-bit data, 19
  extender, 196
  interrupts, STD, 163-166
  lines, data, 11-14, 21
  manufacturers, 277-281
  requesting the, 191-192
  16-bit data, 19

C

Card(s), 205-208
  CPU, 35-38, 46
  interface, 33
  peripheral, 35-38
  Pro-Log 7304 dual UART, 244-258
  Z-80 CPU, 33-37
Clock
  and reset, 267
  signals, 17
Comparators, 60, 61, 73, 81, 197
Control
  bus, 264
  -only
    device(s), 28-29
    peripherals, 29
  signal(s), 14, 15-17, 30-31, 39, 41-42, 43, 61, 65
  generation, 26-27
Controller
  DMA, 191
  MATRIX 7911 stepper-motor, 233-242
  traffic light, 29, 88-92
Converter(s)
  analog-to-digital, 141, 161
  digital-to-analog, 103-108, 141
  interface, analog-to-digital, 141-145

CPU
    cards, 18, 19, 35-38, 41, 46, 63, 154-156, 191
        Z-80, 33-37
    chips, 39, 123, 163
    compatibility, 31-32

D

DAC; see digital-to-analog converter
Data
    bus, 26, 263
        8-bit and 16-bit, 19
    lines, 11-14, 21
    Communications Equipment; see DCE
    /control devices and peripherals, 29
    direction registers, 197
    displays, 109-118
    only devices and peripherals, 29
    Terminal Equipment; see DTE
    transfer timing, 33-35
DCE, 246-247
Decoder(s), 52-60, 61, 197
    circuits, 50-52
    /drivers, 92, 100-101
    x-line to y-line, 52-53
Decoding
    circuitry, address, 148
    using gates for address, 42-50
Designing input ports, 124-130
Device(s)
    addressing, 42
    I/O, 28-30, 44-46
    select pulse, 58-60, 76, 80, 81, 101, 110, 138
Differential
    input, 275
    mode, 216, 219
Digital
    devices, 72
    I/O
        bus, 195-204
        panels, 196
    -to-analog converters, 103-108, 141, 161, 214
DIOB, 195-200 ; see also digital I/O bus
DIOP, 195-204 ; see also digital I/O panels
Direct-memory access, 161, 190-191; see also DMA
Display(s)
    data, 109-118
    LED, 92-102
    multiplexed, 92, 100

Display(s)—cont
    "static," 99
    system, Enlode 214, 205-214
Double buffering, 119
DMA, 11, 190, 191
    controller, 191, 192
    devices, 16
    operation(s), 191-192
    software, 193
    transfers, 191-192, 193
DTE, 246-247
Dual
    DAC interface, 111-112
    UART card, 244-258
Dynamic memory, 25, 26, 191-192

E

Electrical
    fuses, 23-24
    specifications, 267-268
Enlode 214 display systems, 205-214
EPROM; see also erasable programmable ROM
Erasable programmable ROM, 24-25
Expansion, I/O, 16
8-bit
    address, 42, 46, 79, 120
    data bus, 19

F

Flag(s), 134-137, 140, 161, 173, 252-254
    keyboard strobe, 134, 135
    status, 135, 137

G

Gate(s), 60, 123, 165
    AND, 43-45
    for address decoding, 42-50
    NAND, 43-45, 57, 61, 74-75, 141
Glitches, 105, 119, 165
Grounding techniques, 216-217

H

High-priority interrupt, 163, 165, 174-175, 179, 181

I

Input, 173
    devices, 50, 63-65

283

Input—cont
　　ports, 119, 123-159
　　　　designing, 124-130
　　　　memory-mapped, 148-158
Interface(s), 41, 88, 92, 99, 101, 119, 154, 209
　　analog-to-digital converter, 141-145
　　ASCII keyboard, 130-132, 136
　　ATEC 710 thumbwheel switch, 231-233
　　cards, 33, 195-256
　　chips, peripheral, 119
　　compatibility, I/O, 32-33
　　crt, 244
　　DAC, 104
　　dual DAC, 111-112
　　floppy disk, 134, 140
　　input port, 123, 140
　　keyboard, 137-140
　　stepper-motor, 240
Interfacing, 27
　　output port, 81-121
　　parallel, 130
　　serial, 130
Interrupt(s), 161-190
　　and bus control, 265-266
　　and the stack, 184-186
　　bound, 190
　　high-priority, 163, 165, 174-175
　　low-priority, 163, 165, 174-175
　　nonmaskable, 163-164, 165, 175-176
　　operation, basic, 162-163
　　signals, 16, 162
　　software, 181, 193
　　STD bus, 163-166
　　timing, 188-190
　　vectored, 164-165
　　vectors, 28
I/O
　　accumulator, 30-31, 41-42, 60, 66, 79, 120, 156
　　control, 264-265
　　device(s), 28-30, 44-46, 61, 137, 146
　　　　addressing, 41-80
　　　　memory-mapped, 196
　　expansion, 16
　　instructions, 154, 156, 158
　　interface compatibility, 32-33
　　memory-mapped, 30-32, 41-42, 43, 44, 61-65, 66, 79-80, 120, 154, 158
　　software, 193

I/O—cont
　　techniques, memory-mapped, 148

K

Keyboard
　　interface, 137-140
　　　　ASCII, 130-132, 136
　　strobe
　　　　flag, 134, 135
　　　　pulse, 138, 139
Key bounce, 131, 132

L

Latch(es), 82, 84-85, 92, 104, 121
　　chips, 84-87
　　in output ports, 85-87
LED displays, 92-102
Logic tester, 146-147
Low-priority interrupt, 163, 165, 174-175, 181

M

Matrix
　　7911 stepper-motor controller, 233-242
　　Corporation, 259
Memories
　　dynamic, 191-192
　　1-bit "wide," 21
Memory, 20-21, 27
　　address, 11, 61, 63
　　and I/O control, 264-265
　　expansion, 30, 217
　　locations, 19, 27-28
　　mapped
　　　　input ports, 148-158
　　　　I/O, 30-32, 41-42, 44, 61-65, 66, 79-80, 120, 154
　　　　　　device, 196, 231
　　　　　　instructions, 158
　　　　　　techniques, 148
　　　　output ports, 120
　　reference instructions, 31, 154
　　static and dynamic, 25-26
Mode word, 249
Multiplexed, 214
　　displays, 92, 100
　　output, 92
Multiplexer
　　analog input, 145, 273-275
　　channel, 219-221

N

NAND gates, 43-45, 57, 61, 74-75, 141, 165
Nonabsolute addressing, 55, 58
Nonmaskable interrupt, 163-164, 165, 175-176
Nonstandard signals, 39

O

Output(s)
  control, 85
  device, 50
  multiplexed, 92
  port(s), 85-86, 119, 121, 146, 147, 238
    interfacing, 81-121
    latches in, 85-87
    memory-mapped, 120
  three-state, 123-124, 126
  timing, 81-84

P

Parallel
  interfacing, 130
  interrupt-controller card, 167-168, 173
PCI signal, 16, 173-174
PCO signal, 16, 173-174
Peripheral(s), 20, 29, 119
  cards, 35-38, 39
  timing control, 265
PICC, 167-168, 171, 172, 177, 179, 190; see also parallel interrupt-controller card
Port(s), 39, 234, 247
  input, 123-159
  interfacing, output, 81-121
  latches in output, 85-87
  memory-mapped
    input, 148-158
    output, 120
  output, 85-86, 146, 147, 238
Power buses, 263
Priority chain
  input; see PCI
  output; see PCO
Programmable
  devices, 75
  ROM, 23-25
PROM; see programmable ROM

Pseudo-diffential
  input, 273-275
  mode, 216, 219
Pulse(s)
  device-select, 76, 80, 81, 101, 110
  generation, device-select, 58-60
  programming, 24

R

Read-only memories, 23-25
Read/write memory, 25-26
Registers, data direction, 197
Requesting the bus, 191-192
Restart instructions, 166
ROM
  erasable programmable, 24-25
  mask, 23, 24
  programmable, 23-24
RTI-1260, 214-231, 273
R/W memory, static and dynamic, 25-26, 27

S

Sample-and-hold, 217
Serial
  data format, 244-245
  interfacing, 130
  priority, 168, 173-175, 179, 267
Signal(s)
  bus, 12-13
  control, 14, 15-17, 26, 30-31, 39, 41-42, 43
  generation, control, 26-27
  PCI and PCO, 16
Single—
  byte call instruction, 166
  ended
    input, 273
    mode, 216, 219
Software
  ADC, 222-223, 230-231
  DMA, 193
  interrupt, 181, 193
  I/O, 193
  SMC, 242
  traffic light, 93-96
Status flags, 135, 137
STD bus, 3-4, 18
  compatible, 20
  interrupts, 163-166
  processors, 19-20
  Standard, 259-272

STD bus—cont
  what is?, 11-39
  why use?, 18-19
Stepper-motor
  controller, 233-242
  interface, 240
Strobe
  flag, keyboard, 134, 135
  pulse, keyboard, 138, 139
Switch interface, thumbwheel, 231-233

T

Three-state
  devices, 158
  outputs, 123-124, 126
Thumbwheel switch interface, 231-233
Timing, 33, 36-37, 126, 129, 161
  control, peripheral, 265
  data transfer, 33-35
  interrupt, 188-190
  output, 81-84
Traffic light
  controllers, 29, 88-92
  software, 93-96

U

UART card, dual, 244-258
Universal
  Asynchronous Receiver/Transmitter; see UART

Universal—cont
  Synchronous/Asynchronous Receiver/Transmitter; see USART
USART, 244-245, 248-250, 256
Using
  comparators, 60, 73, 81
  decoders, 50-58, 81
  gates for address decoding, 42-50
  PROMs, 66-78, 81

V

Vector, 170
Vectored interrupt, 164-165
Voltage
  analog, 103-104
  input configurations, 273-275

W

Waveforms
  analog, 141
  timing, 33-34, 38

X

X-line decoders, 52-53

Y

Y-line decoders, 52-53

READER SERVICE CARD

To better serve you, the reader, please take a moment to fill out this card, or a copy of it, for us. Not only will you be kept up to date on the Blacksburg Series books, but as an extra bonus, **we will randomly select five cards every month, from all of the cards sent to us during the previous month. The names that are drawn will win, absolutely free, a book from the Blacksburg Continuing Education Series.** Therefore, make sure to indicate your choice in the space provided below. For a complete listing of all the books to choose from, refer to the inside front cover of this book. Please, one card per person. Give everyone a chance.

In order to find out who has won a book in your area, call (703) 953-1861 anytime during the night or weekend. When you do call, an answering machine will let you know the monthly winners. Too good to be true? Just give us a call. Good luck.

If I win, please send me a copy of:

_____

I understand that this book will be sent to me absolutely free, if my card is selected.

For our information, how about telling us a little about yourself. We are interested in your occupation, how and where you normally purchase books and the books that you would like to see in the Blacksburg Series. We are also interested in finding authors for the series, so if you have a book idea, write to The Blacksburg Group, Inc., P.O. Box 242, Blacksburg, VA 24060 and ask for an Author Packet. We are also interested in TRS-80, APPLE, OSI and PET BASIC programs.

My occupation is _____
I buy books through/from _____
Would you buy books through the mail? _____
I'd like to see a book about _____
Name _____
Address _____
City _____
State _____ Zip _____

MAIL TO: BOOKS, BOX 715, BLACKSBURG, VA 24060
!!!!!PLEASE PRINT!!!!!